Technical Communication

Technical

A Practical Guide

Communication

JOSEPH P. DAGHER

Schoolcraft College

PRENTICE-HALL, INC., ENGLEWOOD CLIFFS, NEW JERSEY 07632

Library of Congress Cataloging in Publication Data

Dagher, Joseph P (date)
Technical communication.

Includes bibliographical references and index.
1. Communication of technical information.
2. Technical writing. I. Title.
T10.5.D33 808'.066 77-12116
ISBN 0-13-898247-3

Technical Communication

Joseph P. Dagher

Printed in the United States of America

10 9 8 7 6

Prentice-Hall International, Inc., *London*
Prentice-Hall of Australia Pty. Limited, *Sydney*
Prentice-Hall of Canada, Ltd., *Toronto*
Prentice-Hall of India Private Limited, *New Delhi*
Prentice-Hall of Japan, Inc., *Tokyo*
Prentice-Hall of Southeast Asia Pte. Ltd., *Singapore*
Whitehall Books Limited, *Wellington, New Zealand*

To my family:
Abdo and Hesyne, my parents,
and Ann, Vic, Helen, Fred.
Also in sincere gratitude
to Dr. Jack Newton, my friend.

Contents

Preface

Technical communication is today's most common form of human interaction. Most people live a lifetime without delivering more than one or two formal addresses or writing a single novel, but few live through a day without talking or writing to someone about a commercial, industrial, scientific, or other technical subject: the causes of inflation, how to repair something, how or why something behaves as it does, or how to cook a favorite dish.

When you leave school and go to work in a department store, hospital, factory, or office, you will spend a large part of each day in some form of technical communication. You will need to know how to write effective business letters and job résumés or to give instructions. You will have to make inquiries, place orders, file claims, explain how something should or shouldn't be done, define the meaning of a technical term, or express value judgments. All of these are discussed and illustrated in the pages that follow.

This book will show you how to speak and write more purposefully about the things related to your present and future occupations. It will instruct you in practical principles of technical communication and give you practice in applying them. In helping you improve your speaking and writing about industrial, commercial, and scientific ideas, it will also help you understand better what others say and write along these lines.

This book includes a number of features and approaches that have proved especially helpful to students in improving their communication skills:

1. This book emphasizes some of the nonverbal devices—such as those applied in the dynamics of human interactions, as well as charts, graphs, drawings, and photographs—that are often as important in technical communication as the words. Visual experiences are especially important today, especially since television and other pictorial media have conditioned us to rely on what we see in interpreting and communicating ideas. Therefore a special effort is made here to help you combine visualization with verbal communication.

2. Since a person can't write an effective communication without understanding sentence and paragraph development, this book discusses both. You will also learn to write a variety of letters, a résumé, and a series of technical reports. A grammar review shows you how to apply principles of grammar and punctuation.

3. Many of the illustrations in this text were written by students. These will help you see that the principles and techniques presented here are within your range of achievement. These student illustrations are good, but not professional; you may be able to do better.

4. Research, both on the job and in libraries, plays an important part in technical communication. You will be given practical tips on how to do your research, check the reliability of your sources, gather your information and assemble it into the required formats.

5. Learning to write better helps you read better. When you know the route to some distant city before you start, you find the trip easier; this is also true in reading. When you recognize the road signs—certain kinds of ideas and why they are included—you are better able to interpret what you are reading. You can also distinguish more readily what you should read closely from what you may skim over lightly, and thus you will increase your reading speed.

6. Written communication begins with thoughts conceived as spoken words. You will learn the important similarities between speaking and writing in technical communication. You will see how to apply the principles of human dynamics in speaking and listening on the job. You will learn to say what you think worth saying, not only with written words; you will also be given an opportunity to apply these principles as a speaker before the class and as a reader and listener.

Each year people in industry fabricate not only millions and billions of tons of metal, glass, lumber, brick, cloth, and finished products such as automobiles but also millions and billions of words. An industry's most costly products, its most valuable possessions, may be the words in its file cabinet—the descriptions of production processes and the evaluations of suppliers and customers that are essential for developing, manufacturing, selling, and distributing a product. It has been pointed out that if all the paper containing the words required for its planning, design, construction, and promotion were loaded into a 747 jet plane, it would be too heavy to lift off the ground. Some day you may want a job requiring you to write some of this valuable material. Your employers will be justified in insisting that you produce communications as high in quality and as competitive in cost as their other main products. Training you to do just that is the purpose of this book.

Acknowlegments

I wish to express my deeply felt gratitude to all who helped me in one way or another with this book. For their helpful suggestions I thank Mary Joan Woods, C. Jeriel Howard, and Jack Gilbert. I am also grateful to the dedicated members of the publisher's staff: to Hilda Tauber for her diligent editorial help and advice, to Bill Oliver for his friendly persuasion and wise counsel, and to Sue Herrick for her enthusiastic encouragement. I sincerely appreciate the help of all my students—those whose work is used for illustration and those who contributed in other ways. From all of them I learned as much as I taught.

1

What is technical communication?

technical vs. general communication . . . the main elements . . . language: oral and written, verbal and nonverbal . . . the communicator's purposes: to inform and to persuade . . . reader relationship: establishing contact;. choice of viewpoint and tone

If you were asked to build a gazebo, before starting you would have to know what one is. Similarly, before beginning to build skill in technical communication, you have to know what it is and how it differs from other types of communication.

The main difference between communication in general and technical communication is in their message. This can be easily seen from the following definitions:

> General communication is the use of effective language to express a message to achieve a predetermined purpose.
>
> Technical communication is the use of effective language to express a *commercial, industrial, or scientific* message to achieve a predetermined purpose.

Technical communication, like any other kind, has three main elements: (1) language, (2) message, and (3) predetermined purpose. These may be defined as follows:

What is technical communication?

1. *Language* is a series of verbal and nonverbal sense embodiments of ideas.
2. *Message* is the contents, the ideas expressed.
3. *Predetermined purpose* is the response wanted from the listener or reader.

Clearly, technical communication is more than just transmitting or expressing ideas. It is a controlled use of language to convey ideas for a certain purpose. To communicate successfully, you must keep these elements of the technical communication process in mind. A poor technical communication will have one or more of them incorrectly developed, and a good one will develop all three effectively.

WHAT IS LANGUAGE?

Language consists of signs and symbols that express ideas in a form that can be sensed. It is a series of related words, numbers, and marks that can be heard or seen by the receiver of the message. Human beings communicate through their senses; therefore, language must first stimulate one or more of the five senses. Even ideas such as *space, nothing, honor, honesty, soul, love,* or *hate,* to be expressed and understood, must first take some form that can be heard or seen.

Language can be verbal or nonverbal. Verbal language consists of words; everything else you use to express what you have to say is nonverbal language. The paragraph you are now reading has both verbal and nonverbal elements. Along with the words, it contains capitalization, punctuation, and spacing between words. It also has paragraph indentation and an Arabic numeral. These are often-used nonverbal devices. Later, in Chapter 8, graphics, charts, and drawings are explained at length. These graphic illustrations are essential language elements in technical communication.

Human beings have devised many ingenius nonverbal forms of language for expressing their thoughts and feelings. Some of them unleash us from the limitations of time and space, enabling other people who lived many years ago in far-off places to continue communicating today. For example, through their stone and marble figures the sculptors of the Italian Renaissance transmit to our generation their ideas about beauty and human form. The table on the next page lists some kinds of sense language.

Oral vs. Written Language

It is important that you clearly understand the differences between oral and written language. When you think verbally, you think orally. That is, you not only hear the words, but you hear a voice saying the words, and not just in a monotone, but expressively. If you pay close attention to what happens while you think, you will detect the nonverbal elements of oral expression. Looking at a mirror as you think, you will see various changes in your facial expressions while you express your thoughts to yourself. You almost will be

NONVERBAL LANGUAGE ELEMENTS

Sight	*Sound*	*Smell*	*Touch*	*Taste*
Packaging	Sirens	Perfumes	Slap	Candy
Designs	Foghorns	Incense	Kiss	Holiday foods
Fashions	Whistles	Chemical gases	Hug	Ritual foods
Colors	Auto horns	Food aroma	Handshake	Sacramental bread and wine
Candles	Music		Braille	
Directional signals	Drum beat			
Traffic lights	Humming			
Statues	Morse code			
Flags				
Paintings				
Trademarks				
Totem poles				
Smoke signals				
Uniforms				

able to hear your voice intonations. Think the word "laugh" and then the word "murder" and try to detect changes in your voice and facial expressions. Now, read silently the rest of the words on this page and notice how you use your inner voice to help you interpret what you read. Reading is a kind of thinking.

As a technical writer, you have to know how to translate your oral thoughts into written language to communicate them effectively. To do this, you should learn as much as you can about the techniques of spoken language and the extent to which they can be translated into writing.

Sometimes spoken language is more successful than written, sometimes the reverse. To express ideas a writer mainly uses words, punctuation, capitalization, spacing between words, and paragraphs. Along with words, a speaker can also use vocal intonations, facial expressions, and limb gestures.

Speakers manipulate the meanings of language by varying the stress, pitch, volume, and pace of their voices. Let us look briefly at these devices.

Expressive Devices

Stress is an emphasis of certain syllables or words. For example, in answering the question, "Did you do your report?" you can vary the meanings of the words merely by shifting the stress. Notice how the meanings of the following answers differ, even though the same words are used:

Yes, I did. (It is a fact that I did.)

Yes, *I* did. (Maybe others didn't but I did.)

Yes, I *did*. (How dare you doubt that I did?)

Pitch is highness or lowness of sound in the musical scale. It is the high or low frequency of voice in which we express different meanings. An employee can say, "The boss thinks I'm not honest," in a high-pitched voice to mean that the boss is absurd. She can say it in a low pitch to convey the hopelessness of contradicting the boss. Or she can say it in a normal pitch to report the situation objectively.

Volume is loudness or quietness. When we shout, we express different emotional meanings than when we speak the same words softly.

Pace is speed. When we say something rapidly, an element of concern, excitement, fear, or perhaps even hysteria is added to the meanings of the words.

Facial expressions are very important in conveying meanings. We wrinkle our foreheads to show doubt or inquisitiveness. We expand our cheeks and turn the ends of our lips up to show happiness or friendliness. To show displeasure or anger we flatten our cheeks or draw them in and turn the ends of our lips down. We open our eyes very wide in surprise or astonishment, or we narrow them in anger and suspicion.

Head movements are used to change or shade the meanings of almost every word we use. We lean the head one way or another to tell how we feel about an idea. We shake it vigorously or gently from side to side or up and down to indicate different degrees of "Yes" or "No." Scratching the head, of course, is used to show doubt or confusion.

Limb gestures are essential for effective oral expression. Pointing to help give directions, extending the hands palms up to ask for something, extending them palms down to reject something, or waving an arm to express "goodbye" are only a few of the many limb gestures we use.

Body movements are often used in spoken language. When speakers walk or bend toward the listeners, for example, they indicate the importance or the confidential meaning of what they intend to say.

Being in direct contact with the listener, a person speaking has other advantages a writer doesn't have. You must realize this when converting your thoughts into written form. As a speaker, you can see from the expressions on your listeners' faces whether or not they understand. If their faces indicate confusion, you will be able to repeat your points. Also, your listeners can stop you and ask you to explain. From your listeners' reactions you can decide how much explanation you have to give and to what extent you may have to amplify it by means of illustration. You can also determine whether your reasons and logical arguments are being accepted.

Advantages of Written Language

Oral communication has distinct disadvantages; that's why you have to know how to communicate by using written language. Most of the important technical ideas you express would be lost and forgotten soon after you said them if you had no way of writing them down for future reference. Oral communication is limited by space and time. People in Europe or Asia do not rely on vocal communication as the most convenient means of expressing routine business messages to people in America. Although they can do so by telephone, they would more than likely use the written word, probably a letter.

With oral communication alone, we would not have been able to make as much technological progress, because the spoken word is restricted by time. Written language enables such great technological minds as those of Copernicus, Galileo, Sir Isaac Newton, and Marconi to continue to tell us their thoughts about certain things. They influence our lives today more than they influenced the people of their own times.

Human beings have invented many wonders, from the wheel and the printing press through miracle drugs and the laser beam. How many of these great things would have become real today without the aid of technical writing? Each new inventor stands "upon the shoulders of giants"—the thoughts and work of his predecessors, expressed in writing. Today, stacks upon stacks of technical writings are available in our libraries. Some of us in our work will draw on this warehouse of technical knowledge; some, by writing reports or books, may add to it so that others may use it in making their own discoveries and inventions.

Written language, like oral, consists of nonverbal as well as verbal elements. Listed below are a few of the many nonverbal signs and symbols you may use to help translate orally conceived ideas into written form.

- ⊖ or ⊕ Earth
- ♂ or ♁ male
- ♀ female
- © copyright
- ℅ care of
- ℀ account of
- @ at
- ¢ cent
- ℞ take (from Latin Recipe)
- lb pound
- # number
- % percent
- ∠ angle

- π pi
- ° degree
- ′ minute
- ″ second
- HP horsepower
- ○ moon
- ☿ Mercury
- + plus
- − minus
- ± plus or minus
- × multiplied by
- ÷ divided by
- = equal to

THE COMMUNICATOR'S PURPOSES

The predetermined purpose of any technical communication is the response the person writing or speaking wants from the intended receiver. When you say or write something, you usually want the listener or reader to react in some way, and your message is carefully shaped to win that reaction. Experience shows that it seldom works otherwise. When you communicate about technical subjects, you should decide before you sit at the typewriter or pick up a pen the kind of response you want from the receiver. This enables you to determine the kinds of words to use, the amount of a certain kind of information you will need, and the extent of development required to achieve the intended reaction from the receiver.

A technical communication is usually designed to achieve either of two predetermined purposes: to inform or to persuade. Each purpose aims to arouse specific responses:

Purpose	*Reader Response*
To inform	Understanding and accepting new information
To persuade	Abandoning or changing an established conviction and, if required, doing something as a result

Purpose: To Inform

When you know before you start to speak or write that you intend to inform, you know you will have to offer mainly facts and logical inferences derived from reliable sources. Any opinions you put forth will have to be those of people whom the reader or listener regards as authorities. The qualification of the authors and others from whom you secure your information will play an important role. Therefore, when you think it necessary, you should not hesitate to tell your receivers where you derived your information.

Sometimes, especially when they are not experts, you may have to provide your readers or listeners with a little entertainment to keep them interested as you inform them. Graphs, charts, drawings, photographs, and other visual aids may be helpful here.

It is easy to see from its contents that the main purpose of the following statement is to inform. The new information it contains is presented so that almost anyone can understand and accept it. The writer is intentionally aiming at the layman, not the expert. Notice how the new information is related to what the reader understands, and care is taken to define any terms that may cause confusion.

> During the Pleistocene, the Great Ice Age, which began about a million years ago, mountain glaciers formed on all continents. In some places in North America and Eurasia, they were several thousand feet thick. Almost a third of the present land area was ice-covered. Great glaciers even today cover almost a tenth of the land. This glacier area tells us what the conditions were like during the Great Ice Age. It is possible, therefore, to reconstruct the extent and general nature of these glaciers.
>
> Swiss peasants living in the 19th century believed that the Alps glaciers had been much larger, extending farther down the mountain sides. They noticed that *erratics* (boulders) were being carried by today's glaciers; therefore, they concluded that the many other boulders now in the valleys must have been transported by other galciers long ago. Today, the story of the Great Ice Age is being unraveled by the field observations and new techniques of specialists in many fields of study.
>
> Accumulating field observations, new theories, improved techniques, and worldwide studies of existing glaciers are bringing a clearer under-

> standing of the Pleistocene epoch of the earth's history. This period was a time of spectacular development, of drastic changes in the earth's climate, sea level, plant and animal life, and of the great glacial ice sheets. It caused many changes, and without some of them, human life could not have survived. The environment in which we now live has been largely brought about by the events occurring during the Great Ice Age.[1]

When your purpose is to inform, your main goal will be to provide the receivers with new information they can understand and accept, but you will not try to persuade them to abandon convictions. When you aim at getting them to discard already established beliefs for yours, your intended purpose will be to persuade, not to inform. Consequently, you will need different kinds of ideas, and you may have to express them in a different way.

Purpose: To Persuade

Before deciding to persuade, you should first be sure that you have an adversary—someone disagreeing with you about your conclusion. When you try to convince a person who likes to buy only large, expensive cars that doing so is a waste of money, you can be reasonably sure that he or she disagrees with you and will argue the issue.

When your purpose is to persuade people to change their convictions by accepting yours, the communication will have to contain certain kinds of ideas. You will have to present reliable evidence supporting and proving your argument and showing logically how you arrived at your conclusions. You will offer the testimony of qualified authorities to support your statements.

Persuasion may be aimed at getting others either to change convictions or to do something as a result of changing them. After persuading the buyer of limousines that doing so is a waste of money, you may try getting him or her to go out to buy a smaller car instead. If you succeed, you have persuaded the person to make an intellectual change and to physically do something as a result.

To persuade listeners or readers, especially when they are subjectively attached to their beliefs, it often is necessary to insert some emotional appeal. It is very difficult to change people's convictions about certain organizations to which they belong, such as labor unions or political parties, or about products they buy and use, without relying upon emotional appeal as much as upon logic. Most advertising appeals to emotions in attempting to persuade people to buy a product.

The following example aims to persuade its readers that licensing auto mechanics is not an effective means to insure competence and honesty in repairing automobiles in Kentucky. Notice how the information is cast

---

[1] U.S. Department of the Interior, "The Great Ice Age" (Washington, D.C.: U.S. Government Printing Office, 1969).

in the form of reasons why licensing will not guarantee skill and ethics. The writer also uses reliable authorities—the State Attorney General and the Kentucky Consumer Protection Division—in support of the argument. Notice that the whole article is aimed at an adversary. Here the adversary is someone who firmly believes that mechanics in Kentucky should be licensed.

> Kentucky legislators think they can reduce the number of dishonest and incompetent auto mechanics by requiring them to secure a license to operate a garage. Requiring them to secure permits or licenses is not in the best interest of automobile owners because no law or license can guarantee that a mechanic is competent or honest.
>
> Plumbers, electricians, doctors, and lawyers, for example, are certified and licensed. Some are required to take extensive training, to serve long apprenticeships, and to pass difficult examinations. As anyone who reads the papers today knows, legal action against malpractice and fraud is still common.
>
> Auto agencies with service departments are now required to secure licenses to conduct business in Kentucky. The Consumer Protection Division does everything it can to protect people, but as Frank Malley, State Attorney General, says, "Mediating complaints involving auto repairs is probably the most difficult encountered. . . ." He further states in the January 24, 1976, issue of the *Stouteville Record*, "In cases of incompetence, failure to make proper repairs, excessive charges, repairs without permission, etc., it is difficult—if not impossible—to establish that the dealer or the agent made fraudulent representations."
>
> Trade unions and professional organizations are better able than the State of Kentucky to police and insure the competence and moral integrity of their members. Instead of demanding a license, the state government should encourage and assist mechanics to form a strong, ethical trade organization which can regulate its participants.

Both purposes, to inform and to persuade, will be explained and illustrated further as we study various forms of technical communication throughout the book.

READER RELATIONSHIP

The choice of specific language and the way it is used in a particular technical communication depend upon another vital consideration: the relationship to the intended receiver.

Every communication, technical or otherwise, is between a sender and receiver; it is a process, a system that involves both. When the receiver does not respond in some way to the sender, there is no communication. When the receiver responds as the sender intended, the communication is a success.

Reader relationship

For a communication to be successful, therefore, technical writers or speakers must know enough about the background and interests of their receivers to establish contact and keep it until the intended purpose is achieved. Knowledge of important characteristics of the receiver, such as age, occupational experience, formal education and training, and organizational membership, will help you decide which ideas to express, how to express them, and, just as important, which to leave out.

When the writer or speaker knows the receiver's occupational and educational experience in a certain technical field, he or she can decide what kinds of words and technical terms to use, how much explanation is needed, and how deeply to analyze a complex subject.

The extent of the reader's influence on what writers say and how they say it is clearly illustrated by the following examples defining laser beams. The first paragraph is intended for readers with a knowledge of light and electricity, perhaps with some expertise in the field. Therefore the paragraph is written tightly, without definition of technical terms, using a vocabulary appropriate for educated readers.

> Laser beams are discrete frequencies of highly amplified emission of coherent radiations of light. They are converted from incident electromagnetic fields of mixed light frequencies to discrete ones by extensively increasing the radiation intensity. Laser beams, unlike sunlight and artificially created light, have an extremely narrow frequency range; consequently, they reflect a single color and radiate a great distance from their sources, their edges remaining almost parallel.

In the following paragraph the same ideas are adapted to the background of a general reader, not an expert. Notice how the writer uses a more familiar vocabulary, defines the technical terms, illustrates complex principles of physics, and avoids discussing technical ideas beyond the reader's understanding.

> Laser beams are separate but closely related emissions of light rays greatly increased in strength. They are formed by changing mixed fields of electricity, such as those in sunlight and electric lights, into individual, unmixed ones by stimulating the light rays. The beam from a laser differs from light from the sun or a bulb in that its beam is very narrow. Laser beams are so narrow that a beam one-half inch wide spreads to only about three inches a mile away from its source. An artificially created beam of electric light, such as one from a flashlight, spreads several feet at only a short distance.

Choice of Viewpoint

To help you establish and maintain the needed relationship with your receiver throughout a technical communication, you have to decide in advance two important things: viewpoint and tone. *The viewpoint is the way you decide to observe and think about your subject, and the tone is the way you intend to regard your receiver or type of receiver.*

When you decide in advance how you should see or think about your subject, you will be able to determine the ideas suitable for helping the intended listener or reader share that viewpoint. Knowing your receiver's background, you will be able to decide which ideas about your subject are too complex, too simple, or just right for the kind of relationship you need. You will know how intensely or how superficially to observe or think. You will know, also, how much your personal biases or feelings should be allowed to influence what you observe.

When communicating with persons you regard as experts in a certain field, you will want to view your subject from an *objective viewpoint*. You will take great care to observe the thing or idea you are viewing accurately and realistically. You will do everything you can to keep personal feelings from distorting your perceptions. You will avoid loaded words—those tinted or saturated with your emotions.

The following discussion of sound resulted from the writer's decision to regard it objectively.

> Sound is an undulatory motion of the air or other elastic medium which, falling upon the ear, is capable of producing the sensation of hearing. Any sudden displacement of a part of a continuous elastic medium is transmitted through the medium as a pulse of displacement. If the frequency of these alternations lies within the range between 20 and 20,000 per second, the ear detects the sound and the brain recognizes and identifies it.

The *semiobjective viewpoint* is used in most of the popular magazines dealing with crafts and trades in this country. You should use it when your intended readers or listeners can understand a technical subject fairly well, but may not be trained or educated enough to understand it as experts. You may allow some of your personal feelings to color and shade your ideas, stimulating the receiver's interest. However, you must be careful not to permit too much emotion to distort the accuracy of your message or reduce its effectiveness, so that you fail to inform or persuade.

Here is another paragraph about sound, this time written from the semiobjective viewpoint. Notice how the plain words and the picturesque illustrations reflect—and invite the reader to share—the writer's personal feelings about the subject.

> Sound is made by vibrations resulting from wavelike changes of pressure moving through air. These pressure changes are much like the ripples caused by a pebble thrown into a calm pool of water. The vibrations from the wavelike movement form a pattern of a number of beats per second, in other words, a pulse. When the number of beats in the pulse is between 20 and 20,000 per second, the human ear detects them and the brain identifies the kind they are. Touch lightly the edge of a small bell after tapping it gently, and you will feel these wavelike vibrations. When these vibes are regular, you will hear enjoyable harmony. When they are irregular, however, you may hear irritating noise.

Reader relationship

The *subjective viewpoint* is used when you do not want to deal with your subject too technically. You may want to give the receiver just a bit of information in an interesting manner, as is often done in writing for lay people having a smattering of knowledge about a subject.

The subjective viewpoint is used in the following paragraph about sound. Notice how much emotional appeal the writer uses.

> Rumble, roar, shriek, rattle, and hiss are only a few of the different vibrations that make the sounds you hear. They are all made by sound waves moving back and forth through the air. Think of these waves as ripples in a pool and imagine a cork riding the surface. The number of times the cork passes a certain spot as it floats back and forth causes the beat or the pulse of the sound waves. When this beat is between 20 and 20,000 per second, your ears will detect it and your brain will be able to identify the kind it is. When these vibes are regular, the sound may be the kind of enjoyable music you like to listen to, perhaps jazz. But when it is irregular it becomes noise, the kind that may make you want to climb a wall.

Choice of Tone

Besides viewpoint, the sender/receiver relationship depends upon tone. Tone reflects the way you regard your reader or listener. It indicates the attitude you take toward him or her as individuals or as types in order to succeed in informing or persuading. Tone is the voice in which you communicate. Almost every time you speak to someone you use a different tone of voice; you choose it quickly and almost automatically. To improve your spoken or written communication, you may need to select your tone more deliberately.

Formal, semiformal, and colloquial are the three tones from which you may choose in technical communication. There are various degrees of each. A written report, for example, may be written in a very formal, fairly formal, or slightly formal tone; this is also true of the semiformal and colloquial.

Once you've decided how you should regard your reader, you will be able to determine the kind of language to use. When you regard readers as experts, you know you may use some technical terms without defining them. Therefore, you express complex ideas with precise, sophisticated words that do not sacrifice accuracy for easier understanding.

The somewhat formal tone in the paragraph below is aimed at the reader having some knowledge of dentistry. Etiologic agents, periodontal disease, dental plaque, gingival, and interproximal are all terms a person involved in dentistry would understand without explanation. Also notice the complexity of the other words. You can easily see how impersonal the *formal tone* is.

> Oral hygiene is effective in the prevention and control of periodontal disease. Successful application necessitates the removal of the etiologic agents causing it. Dental plaque causes gingival inflammation when left in the interproximal areas. Gingivitis develops within three weeks after cessation of oral hygiene. However, this inflammation is reversible within three weeks after good oral hygiene is initiated and continued.

The next illustration shows how a writer or speaker uses plainer words when communicating with readers having less technical knowledge. This kind of receiver is able to apply the skills involved in a certain field. Unlike the expert, however, he or she may not be able to deal with the related theory. Therefore, fewer technical terms are used, and they are defined further. Notice how the terms "etiologic agent" and "dental plaque" are carefully explained and how the phrases "proper tooth brushing" and "gum disease" are substituted for the more technical dental terminology. This results in a *semiformal tone* that is somewhat less impersonal than the formal tone.

> Proper tooth brushing and gum care is effective in the prevention and control of gum disease. To brush teeth properly, a person must remove the etiologic agent, the cause of the disease. The main cause of gum disease is dental plaque, a hard substance formed by bacteria. Not brushing allows this plaque to remain on and between the teeth and gums, causing irritation and infection. After three weeks of not brushing, the gums will begin to bleed. This situation can be reversed, however, by starting and continuing to carefully brush the teeth and gums.

The next paragraph expresses the same ideas in a *colloquial tone*. To maintain the right relationship with the reader, the writer omits the dental terms and substitutes simple explanations for them. Bacteria become germs, and the other words chosen are the plain, everyday kind you use when talking with close friends. The very personal relationship with the reader is achieved also by using short, direct sentences and addressing the reader as "you."

> Brushing your teeth the right way can keep you from getting gum disease. To brush them right, you have to get the germs off them. These germs put a hard deposit with sharp edges on and between the teeth. These sharp edges cut into your gums, making them bleed and become infected. After about three weeks of brushing your teeth the right way, your gums will become pink again and won't bleed. This will help you keep your teeth a lot longer, and you won't have trouble with your gums anymore.

Here are some examples showing how words reflect different tones:

Formal	*Semiformal*	*Colloquial*
intoxicated	drunk	tipsy
demented	insane	crazy
chaos	disorder	mess
logical	reasonable	holds water
inquire	ask	snoop
demolish	destroy	put an end to
renovate	renew	remake
extricate	release	set free
exhausted	tired	tuckered out
debase	lower	take down a peg

Reader relationship

Technical communication, then, always aims at arousing a certain response from an intended receiver or type of receiver. To achieve this aim, the writer or speaker must select a viewpoint, tone, and vocabulary that can establish the receiver-relationship needed to inform or to persuade. Also crucial to receiver response are the kinds of ideas the message contains; these are the subject of the next chapter.

Application 1–1

After examining the clothing someone is wearing, write a paragraph explaining what it says nonverbally about the wearer.

Application 1–2

Find some quiet spot and concentrate on something you feel strongly about: war, abortion, taxes, examinations, a person you love or hate, etc. As you are thinking, try to become conscious of the ways you are expressing your ideas to yourself nonverbally with facial expressions, finger and hand movements, and changes in the sound of your voice. Make a list of these nonverbal language elements for discussion in class.

Application 1–3

Select some reading material, read it to yourself, and try to detect nonverbal language as you did in the preceding assignment. Read the same passage out loud where no one can hear you and try to detect voice changes caused by your interpretation of what you are reading. Make a list of these nonverbal language elements for class discussion.

Application 1–4

List:

1. At least five ways human beings communicate by sense language.
2. At least five ways animals communicate by sense language.
3. Five ways in which human beings communicate nonverbally.
4. Several thoughts a person expresses by his or her possessions: cars, jewelry, furniture, house, etc.

Application 1–5

Examine a bottle or some other kind of packaging. Now write a paragraph explaining what its shape, design, and ornamentation say about the product and the manufacturer.

Application 1–6

Identify the writer's purpose for each of the following statements:

1. The railroads in this country should be owned and operated by the U.S. Interstate Commerce Commission, not by private enterprise.
2. While in his automobile, the President of the United States is able to keep in contact with his office at all times.
3. Installment buying has been proved a stimulant to the sale of televisions and automobiles.
4. Credit cards should be issued only to those who have never had credit problems.
5. Erosion in a desert area is not easy to detect.
6. The erection and sale of condominiums is booming.
7. All mechanics should be required to secure a state permit before opening a garage.

8. The sale of merchandise in Detroit groceries increased as much as 10 percent over last year.
9. A well-built house usually has a concrete foundation reinforced with steel mesh.
10. Modern toys stunt a child's imagination.

Application 1–7

Select an idea from some kind of work you have done or from some occupation in which you are interested and write a short composition to inform. Next, select another idea about the same subject and write another composition intended to persuade. Be ready to explain why you think it will achieve the purpose you intend. Also, be prepared to present either composition orally before the class.

Application 1–8

Write an informing paragraph about store windows or store displays. Next, rewrite the paragraph so that it persuades the reader. Be able to explain your paragraphs. Your instructor may ask you to present the ideas in your paragraphs to the class orally.

Application 1–9

Select a television program that you enjoy; turn down the volume, leaving only a silent picture. Closely watch the nonverbal language elements the actors use to convey the precise shade of meaning: facial expressions, limb gestures, body movements, etc., and make a list of them. If you can, try to record how these things affect the meanings of the words uttered. Next, turn up the volume and listen for voice manipulations that alter the meaning; especially watch the stress, pitch, volume, and pace.

Application 1–10

Be able to tell whether each term below is formal, informal, or colloquial in tone.

1. askew
2. climb
3. cockeyed
4. dignitary
5. ascend
6. diverge
7. line up
8. important person
9. transposition
10. use up
11. climb
12. emphasize
13. drop
14. VIP
15. pad (apartment)
16. coextend
17. descend
18. upside down
19. dwelling
20. accentuate
21. angle
22. highlight
23. invert
24. deplete

2

The technical message: facts, inferences, opinions

fact, inference, opinion—what they are . . . how they differ . . . detecting them . . . writing them as you intend them to be understood

The technical message, like any spoken or written communication, consists of facts, inferences, and opinions. It is essential that you know what these are and how they differ. You must know also how to express them as you want your listeners and readers to understand what you intend each to be —a fact, an inference, or an opinion.

Not only technicians but everyone should know what facts, inferences, and opinions are and how to use them. Many quarrels and painful misunderstandings among family members and fellow workers could thereby be avoided. The following exercise will help you check your own understanding.

Carefully examine the photograph in Figure 2.1, which was taken after an accident. On a sheet of paper, tell whether the statements beneath it are facts, inferences, or opinions. After you have read the chapter, come back and check your conclusions. Be ready to explain your answers in class.

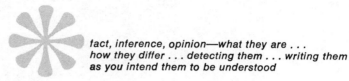

Figure 2.1

1. Car 1 caused the accident.
2. At least two cars are involved in the accident.
3. Car 2 made the tire skid marks.
4. The man on the bicycle witnessed the accident.
5. Car 3 has a front bumper wider than those on Cars 1 and 2.
6. The driver of either Car 1 or 2 was drunk.
7. The accident occurred at an intersection.
8. Traffic is moving around the accident.
9. Car 1 was making a right turn from the wrong lane.
10. No one was seriously hurt in the accident.
11. Car 2 pulled out into the north-south highway too soon.
12. The north-south road is a divided highway.
13. Car 3 is a police car.
14. There is a "one-way" sign at the intersection.
15. The three men standing next to Car 2 were its occupants.
16. A traffic light should be placed at this intersection.
17. Only two cars were involved.
18. A gas station is located near the intersection.
19. The cars involved were driven by men, not women.
20. The accident occurred in the early afternoon.
21. The roads are not wet.
22. One of the cars is a station wagon.
23. Car 2 is leaking something.
24. The police have arrived at the scene of the accident.
25. All of the cars involved were insured.

Facts, inferences, and opinions can be defined in various ways, some quite philosophical and beyond our concern here. We are mainly interested in the technical or scientific definitions of these terms.

WHAT IS A FACT?

A fact is something known to be proved true. "Proved true" means verified to exist or happen at a certain time or place. We verify things to be true with our senses, by seeing, hearing, smelling, feeling, and/or tasting them. True, you can't prove every fact you communicate yourself. You can rely, however, on the statements of reliable, qualified sources for proof. These sources may be living authorities or books and other printed material.

People you can trust because they have had the needed education, training, and occupational experience are good sources for verification of facts related to their specialities. Many of these people are recognized as authorities because of their achievements: discoveries, inventions, books and articles published, and other accomplishments. When you think your listeners or readers may doubt or challenge a fact because they do not know your source's qualifications, it's your responsibility to give them the necessary background.

Here are some facts most people would accept without additional verification:

1. Some human illnesses are hereditary.
2. The first human being to step on the moon was an American.
3. Some mosquitoes communicate malaria.
4. Water is a combination of hydrogen and oxygen.
5. The Earth revolves around the sun.

All of these ideas are accepted as facts because each has been proved true by someone.

Problems arise in identifying facts because our senses are not always as accurate as we would like them to be. Also, our emotions often blur the things we sense. Sometimes even a scientist may think he detects something, ask another expert to verify it, and learn that he was wrong. A fact, therefore, must be provable by another qualified source, another book or person recognized as reliable.

To illustrate how our senses play tricks on us, the students in a writing class were asked to jot down what they thought the ink blot in Figure 2.2 represented. Their answers are listed next to the blot. Would you agree with all their answers?

1. butterfly
2. bug with broken wings
3. crab
4. spider
5. black cat
6. candle holder
7. burned tree stump
8. bottle opener
9. antennas
10. devil or ghost

Figure 2.2

Our senses are not able to verify some facts; consequently, we invent and rely on instruments to help us. Without microscopes, telescopes, cameras, and other devices we wouldn't know that many things even exist. With the development of new instruments and techniques, still more things are being proved as facts. Do you think you could prove that microbes are facts without first looking into a microscope at them?

When you think your reader or listener may challenge the verifiability of your facts, you should be able to present the testimony of at least one more qualified person to support them. Our minds often play tricks to let us sense only what we want to sense, to believe only what we like to defend, or simply to see what we expect or what most readily makes sense rather than what actually is there. The accompanying drawing by M. C. Escher (Figure 2.3) shows how our expectations can trick us into accepting unwarranted conclusions. Can you detect the evidence that your mind disregards in rushing to see this as a workable mill?

Figure 2.3
M. C. Escher, "Waterfall," 1961 (Escher Foundation, Haags Gemeentemuseum, The Hague).

The problem is that the drawing contains contradictions. Water can't flow uphill, nor can it flow from a canal over a waterfall to a lower level of the same canal if the canal actually is all on the same level. Did you detect that the two towers are the same height even though the one on the left is a story higher than the one on the right?

Is the drawing in Figure 2.4 a possible triangle? Get all the evidence before you decide.

The deliberate trickery in Escher's drawings should alert us to the possibility that we may often be mistaken about what we see or think we see. It is easy to understand why, when necessary, facts should be verified by reliable evidence from two or more qualified sources.

Certain limitations apply to statements of fact, making them less flexible than inferences and opinions. Writers or speakers must use them carefully in order to achieve the desired response from the intended receiver.

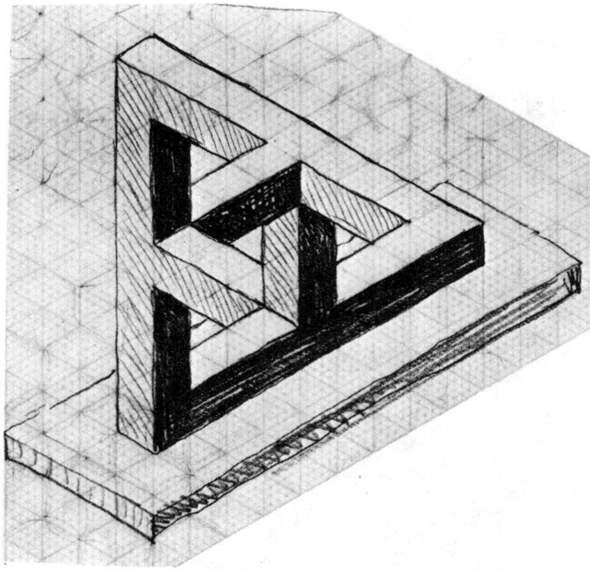

***Figure* 2.4**
M. C. Escher, Perspective drawing of a triangle (Escher Foundation, Haags Gemeentemuseum, The Hague).

Facts are limited by time. They may be true in present time or past time, but not future time. Things and events of the future cannot be facts, cannot be proved true, even though the laws of probability clearly indicate that they might be someday. "Automobiles were equipped with runningboards in 1925" can easily be verified as a fact. "Few automobiles have runningboards today" can also be verified. We can't verify yet that "Automobiles will have runningboards again someday."

Facts cannot be proved true when related to prehistoric times—the period before human beings learned to record evidence. Facts, therefore, are limited to the period during which humans have been able to record what they saw, heard, smelled, felt, and tasted.

Facts are further limited by place. What is a fact in one place may not be in another. That bicycles outnumber automobiles can be verified as a fact in South Korea, but in the United States the opposite is more likely to be a fact.

Whether or not your receiver accepts what you say as a fact often depends greatly upon your supporting evidence. When your purpose is to persuade, especially, you must always test the soundness of your proof. Check to be sure that your sources of evidence are qualified—that they have the needed training and experience and are recognized as authorities by whoever might question or challenge your statements.

People speaking or writing don't have to prove every fact they use. Many of the common facts used in our communications are readily accepted as facts. The people with whom you are communicating will not demand proof that automobiles have internal combustion engines that burn gasoline nor that people in Mexico speak a language derived from Spanish.

Only those facts that you think the listener or reader will question or challenge need proof. Any important idea that you think will be doubted should be supported by reliable evidence or by references to other persons or books that verify your statements.

WHAT IS AN INFERENCE?

An inference is a conclusion logically derived from reliable evidence but for which there is not enough evidence to prove it to be a fact. Later, when enough supporting evidence is found, it may be proved true, or it may be proved to be an incorrect inference.

The following statements cannot be proved true, even though there is a great amount of supporting evidence for them. They are only inferences.

1. At one time, all the earth's continents were joined.
2. The United States will need many more physicians if the population continues to increase at the present rate.
3. The Great Lakes were formed by glaciers.
4. The price of wheat will rise if there is a bad crop next year.
5. Some stars in our galaxy will shine for another 100 million years.

As you can see from these examples, inferences are not limited by time. They enable the human mind to jump over gaps of unknowns to sound conclusions. Scientists conclude from colors on their instruments that the sun is composed of certain gases. They know that these same gases on earth cause these colors to appear on the same or similar instruments.

When we see skid marks on a highway, we may infer that a driver tried to stop a car or slow it down to avoid an accident. From the shape of an island, a geologist may infer that it resulted from volcanic eruptions. We infer that our neighbors are happy, sad, or angry from their expressions and behavior, even though they may be only pretending.

If people were not able to draw inferences from the reliable evidence they discovered, we would still be living in caves. Most of our great scientific discoveries started out as inferences. The word *theory* is just another word for inference. Darwin's theory of evolution is still an inference. It is a great jump forward in our understanding of man, but still a theory, not a fact. Most of the events it is concerned with happened in prehistoric times, before humans recorded observations and passed them down through the ages to be used as evidence. We have made many inferences about outer space. Our trips to the moon have established some of these inferences as fact, proved others incorrect, and left some still as inferences.

Because inferences cannot be proved true, scientists often disagree about them. Some qualified authorities infer from reliable evidence that gorillas use limb and body movements as symbolic language. Therefore, assuming that human beings evolved from gorillas or apes, they believe that the language we use started this way. Other scientists disagree; they think that not enough evidence exists to prove that monkeys, gorillas, and apes can use symbols to express abstractions—ideas they can't see, hear, feel, smell, or taste. With more reliable evidence, this inference may be proved either true or false.

Unlike a fact, an inference often must rely heavily upon the reliability and qualifications of its source. If your doctor believes that your pulse rate indicates you are ill, you will accept that inference as reliable and follow any remedies that he or she prescribes. But if your trusted auto mechanic told you the same thing, you would pay little attention to it.

An inference is not as tightly limited by time and space as a fact is. Often it is a statement of probability. For example, from statistics about the present rate of increase in business, a manufacturer infers that by a certain future time a company building will have to be expanded. An insurance company's whole existence is based on the law of probability. It bets that you will live probably for another fifty years, and you bet you won't. The one drawing the correct inference is the winner; either the company or your heirs will receive the financial benefits. We infer from facts today what the weather will be tomorrow, and we make plans accordingly.

WHAT IS AN OPINION?

Everyone has the right to an opinion. But this doesn't mean that others must accept and react to it as though it were a fact or a logical inference based on sound evidence.

An opinion is based on personal emotions or attitudes rather than objective, reliable evidence. A person may have a fondness for people from Lebanon and assert that all the inhabitants of that pleasant country are friendly. That is an opinion based on personal experiences and attitudes. The individual has the right to hold and defend that opinion and to act as a result of believing it—but if he or she expects others to accept it as a fact and to respond to it as though it were, problems in communication will result.

Here are some other opinions:

1. Most fat people are good-natured and easygoing.
2. Henry Fonda is truly a great actor.
3. Cats are easier to keep as house pets than dogs.
4. Earth has been visited by people from other planets.
5. To have great wealth is to have power.

What is an opinion?

You can easily see how these judgments are based mainly on personal tastes or attitudes.

Unlike facts and inferences, opinions do not exist independent of an individual. When people die, opinions they have die with them. Other people may have similar ones, but they are based on their own tastes and attitudes and so are not exactly the same as those of the person who died. Facts and inferences, however, exist independently. When people die, the fact that the Civil War was fought between the states will continue to exist. The inference that life exists on other planets will continue to be accepted as probable. Both facts and inferences depend upon the reliability of the evidence, not upon personal tastes and attitudes.

All this does not mean, however, that an opinion is never reliable. When your family doctor tells you that you should come in for a physical examination every so often, you accept his opinion as reliable and worth some serious consideration. The doctor's opinion doesn't become an inference until there is more concrete evidence that you need closer attention. Perhaps on one visit the doctor detects a heart sound or change in rhythm leading him or her to infer that a problem exists. This inference is checked in the clinic with more accurate instruments such as an electrocardiogram to determine whether or not it is a fact. It doesn't become a fact until the doctor has enough proof.

Most of the opinions we express are those communicating how we feel about something, what our emotions are. As long as we and our listeners and readers know that's all we intend, good communication will occur. But when we try to make the receiver accept our opinion as a fact, we will run into a communication block. This is also true when we try to present inferences as facts. When we say "That car is a dream," we are exaggerating intentionally to express personal feelings, emotions. If we think the other person may misinterpret this as an assertion of fact or reliable inference, we had better add "I think" or "I feel": "I think that car is a dream." There hardly can be misunderstanding about that statement.

A speaker's or writer's purpose determines whether facts, inferences, or opinions are most important for effective communication. When one is preparing a television commercial, facts are not *always* as important as opinions. When one wants to entertain or to convey an impression, facts and inferences are of little importance. To do either job well, exaggeration and emotions must play a large part in the communication; therefore, opinions will play the biggest role. When the purpose is to inform or to persuade the receiver *logically*, however, the person speaking or writing must use mainly facts and logical inferences.

The statements beneath the ink blot in Figure 2.5 are facts, inferences, and opinions based on the viewers' observations. See if you can tell, from the way each is worded, whether it states a fact, inference, or opinion.

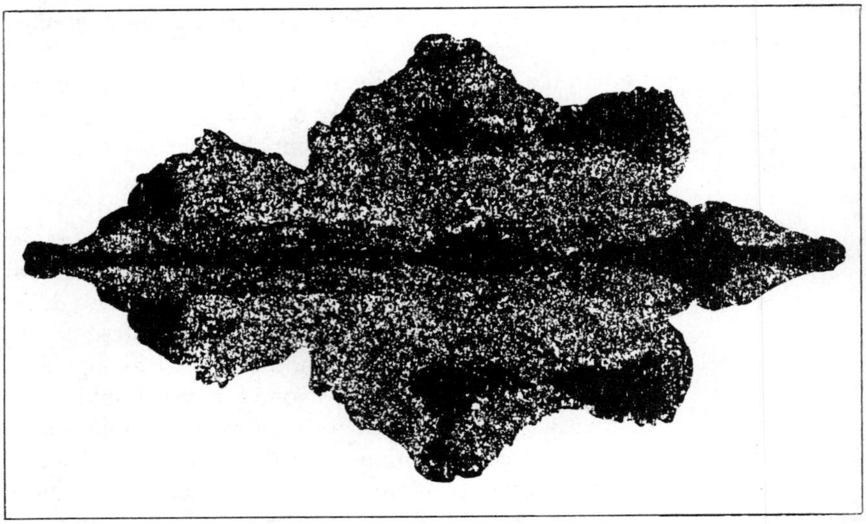

Figure 2.5

1. The blot resembles an alligator.
2. The ink was a liquid when placed on the paper.
3. Each horizontal half of this ink blot is similar to its opposite.
4. It is framed by a rectangular border.
5. Its horizontal dimensions are longer than its vertical ones.
6. Making ink blots like this one is a good way to develop imagination.
7. This blot was made by folding the paper horizontally across the blot.
8. The rectangle is 4½ inches long and 2¾ inches wide.
9. The blot is a beautiful pattern.
10. The blot shape would be different if the paper were folded differently.
11. The ink seems to be distributed rather evenly.
12. It arouses some free associations, if it is allowed to do so.
13. More than a single drop of ink was used in making the blot.
14. A psychologist would refuse this blot to diagnose a patient's problems.
15. The blot is 2½ inches wide across its center.
16. It was fun making this blot.
17. The blot was made with black ink.
18. People see only what they want to in this blot.
19. The distribution of ink is very attractive.
20. Ideas hidden in a person's subconscious may be aroused by an ink blot.

A communicator doesn't have to give the evidence or sources for every statement intended as a fact, inference, or reliable opinion. You must provide support mainly when you expect the receiver to hesitate or refuse to accept your message as you intend it.

Knowing the type of receiver for whom a communication is intended helps the sender determine what must be proved or supported. When communicating with an expert in a certain technical field, you take for

What is an opinion?

granted his acceptance of many of the points that you may have to support and prove for a layman. You have to keep your receiver in mind always, no matter how you are communicating.

Application 2–1

On a separate sheet of paper, indicate whether each of the following is intended as a fact, an inference, or an opinion. Be prepared to explain your answers.

1. On a smooth road, shock absorber pistons move back and forth about 15 times a second.
2. A car's ignition is a rat's nest of wires.
3. Metallurgy is essential knowledge for someone responsible for doing theoretical research in steel.
4. The shop foreman is a considerate person.
5. Diagrams in this technical report show the parts of a door hinge.
6. Metal replacement parts are used in the human body.
7. The generator will be working when you return.
8. This is the world's best generator.
9. The way the material is machined may cause the residual stress.
10. The alloy was hard.
11. Experts in atomic energy know about the hazards of handling plutonium.
12. A break or halt in technological progress is necessary for man to regain respect for nature.
13. Planning and developing aerospace vehicles is a fascinating occupation.
14. In the next one hundred years, man will establish colonies in earth orbit.
15. The automobile industry will employ, in the next few years, people who can perform many new technical operations.
16. Secretaries work with their minds and with their hands.
17. In 1940, space scientists had no rocket capable of launching a satellite.
18. UFO's have flown over and photographed U.S. military bases.
19. Intercontinental ballistic missiles exist today.
20. The letter I received from the purchasing agent was very encouraging.
21. Today, solar cells are used experimentally to recharge batteries.
22. The drawing of the sewing machine is the best.
23. Gasoline pumps are a form of pop art.
24. The rough spots on the metal indicate a need for a different setting on the polisher.
25. A nurse must remain loyal to a doctor always.

Application 2–2

Study the accompanying sketch of an auto accident and write an accident report. Your report might include the names of the streets, the directions of travel, any direction changes, road and weather conditions, witnesses' statements, details describing the collision, extent of damage and injury, and other information about the drivers, passengers, witnesses, along with their addresses.

Now, make a decision about which driver is responsible, give your supporting evidence, and explain how you arrived at your conclusion.

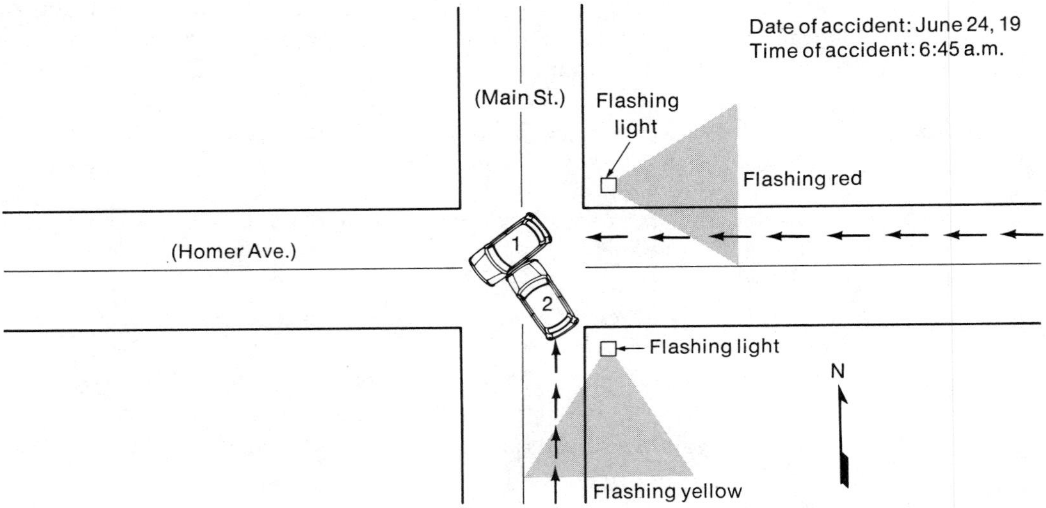

In parentheses before each sentence of your report, print *F*, *I*, or *O* to indicate whether you intend it as a fact, an inference, or an opinion.

Application 2–3 Using your own diagram or a printed accident form available from an insurance company or police station, report an imaginary collision. Be sure to show the position of each vehicle after the accident and the direction in which each was traveling before it occurred.

Application 2–4 Write at least three facts, three inferences, and three opinions based on the information revealed in the accompanying line graph, which shows the relative growth in U.S. population and in availability of fishery products. Also, explain why your opinions are opinions and not facts or inferences.

RELATIVE GROWTH IN U.S. POPULATION AND IN QUANTITY OF FISHERY PRODUCTS AVAILABLE
Includes both edible and industrial fishery products

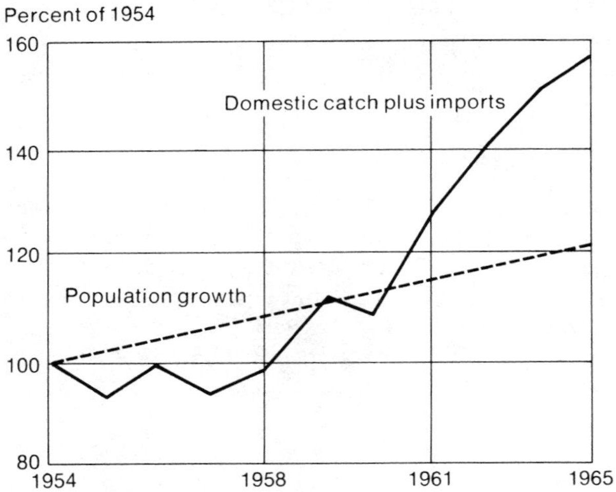

SOURCE: United States Department of the Interior, Bureau of Commercial Fisheries, *The Fisheries of North America* (Washington, D.C.: U.S. Government Printing Office, 1973), p. 12.

Application 2–5 Study the accompanying table, which shows the heart-disease death rates among different age groups of smokers and ex-smokers. From the statistics, write at least five inferences. Under each, present the logical reasoning by which you arrived at the conclusion.

Table 2.1 *Annual death rate per 100,000 from coronary heart disease by age, cigarette-smoking status and number of cigarettes smoked per day, U.S. veterans study*

	45–54		55–64		65–74	
Number smoked per day [a]	*Current cigarette smokers*	*Ex-smokers* [b]	*Current cigarette smokers*	*Ex-smokers* [b]	*Current cigarette smokers*	*Ex-smokers* [b]
1 to 9	195	125	594	432	1,374	1,105
10 to 20	297	133	830	557	1,577	1,260
21 to 39	390	57	912	743	1,701	1,366
40+	502	—	1,101	646	1,955	1,482

[a] This is the current rate of smoking for current cigarette smokers and the maximum rate attained for ex-cigarette smokers.

[b] Ex-smokers who stopped for reasons other than doctor's orders.

SOURCE: United States Department of Health, Education, and Welfare, Public Health Service, *The Health Consequences of Smoking* (Washington, D.C.: U.S. Government Printing Office, 1967), p. 49.

Application 2–6 Based on the accompanying basic bar graph, which shows trends in number of fires and area burned in U.S. National Forests, write three facts, three inferences, and three opinions.

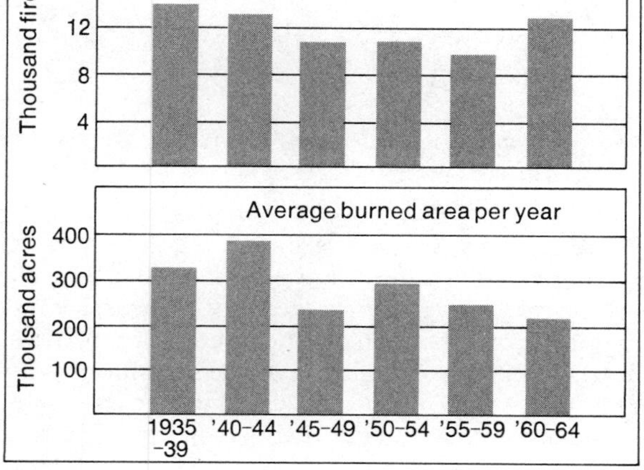

SOURCE: U.S. Department of Agriculture, Forest Service, *Outdoor Recreation in the National Forests* (Washington, D.C.: U.S. Government Printing Office, 1972), p. 88.

Application 2–7 On a piece of paper, identify each of the following as a fact, inference, or opinion. Be prepared to explain your answers.

1. To succeed in life, you must have a college education.
2. Since the horse can't run well on a wet track, he'll lose the race today because it's raining.
3. Large corporations always take advantage of their small competitors.
4. Some international conglomerates have their home offices in the United States.
5. You could tell that the business agent waiting in the lobby was nervous by the many cigarettes he smoked.
6. The wrench was stamped with the words "Made in U.S.A."
7. We should expect a long cold winter because squirrels started collecting nuts early.
8. The chart showed that more accidents occurred at this corner this year than at the intersection of Market and Vine.
9. Judging from the way they dress, the man and woman must be wealthy.
10. The many billboards along our highways show that we do not respect the beauty of nature.
11. Some governments confiscate the property of people who don't pay their taxes.
12. The inhabitants of some European countries have forgotten what it is to be poor.
13. Venice, Italy, is gradually sinking.
14. Puncture-proof tires usually don't explode when pierced by a nail.
15. Young dandelions fried in butter are delicious.
16. If the energy shortage increases, automobile manufacturers will have to increase their diversification.
17. Pigs can survive on roots, acorns, and other things not suitable for human consumption.
18. One of the greatest Chinese dishes is fish head soup.
19. University of Michigan scientists list more than 200 viruses that cause human illness.
20. From the insurance company statistics, the employees figured that they would live to be at least seventy years old.

Application 2–8 Identify the sentences in the paragraphs below as fact, inference, or opinion. All three may not be present in both paragraphs.

(1) More than a century ago, the intimate relationship between glacial ice and the amount of water in the ocean basins was recognized. (2) When the great ice sheet covered vast land areas, sea level was lowered because the normal return of water from the land to the oceans was reduced. (3) Conversely, sea level rose as Ice Age glaciers melted, permitting the melt waters to flow into the ocean. (4) If all the glacial ice on the surface of the earth today should melt, sea level might rise by more than 150 feet.

What is an opinion?

(5) Shoreline fluctuations are also produced through elevation or depression of the land. (6) During times of glaciation the great weight of the ice slowly depressed the earth's mobile crust. (7) Removal of the weight through glacier melting permitted the slow rebound of the crust to its former position of relative equilibrium. (8) Such movement, common in glaciated areas, is best documented in Scandinavia and Finland. (9) Evidence of similar uplift can be observed in the region of the Great Lakes and Lake Champlain, where old shorelines, originally horizontal, are now raised and tilted so that the greatest uplift is to the north.[1]

---

[1] United States Department of the Interior, "The Great Ice Age" (Washington, D.C.: U.S. Government Printing Office, 1969).

3

Speaking and listening on the job

the nonverbal language of human dynamics . . . kinds of technical oral communication: (1) business conversations and (2) group discussions . . . participating as a member or as a chairperson . . . helpful tips for skillful speaking and listening in a technical conversation and in group discussions

Most of the principles we shall learn in this book are related to spoken as well as to written technical communication. To speak or write effectively, you should know the differences between oral and written language and how to translate your thoughts from one to the other. You should know how to use graphics along with words to make what you say clear to your listeners or readers. In both spoken and in written communication you must be able to distinguish facts, inferences, and opinions to express yourself accurately.

Speaking and listening on the job is more than exchanging groups of words and sentences. It also is applied psychology, much of which is achieved by the nonverbal elements discussed in Chapter 1. Speakers use their voice intonations, facial and limb gestures, body movements, their whole personalities to help them say exactly what they mean. It would be wise for you to review Chapter 1 before continuing.

Speaking instead of writing enables people to see and hear their listeners' responses immediately. They don't have to wait several days to learn whether or not they succeeded in their communication. In selling automobiles, styling hair, and almost every other kind of work people use oral communication; direct contact is needed. Even spoken communication by telephone is often better than written, because the speaker can make immediate changes and adjustments to achieve the required mutual relationship.

Often technical communications take place during conversations, on the spur of the moment. Other types, however, must be planned and scheduled. In this text, therefore, we shall study the various kinds of oral communications used in business and industry under two general headings: (1) business or technical conversations and (2) group discussions.

BUSINESS CONVERSATIONS

Business conversations are spontaneous communications between two or more people. When you take part in one, you don't have to be recognized before speaking, and you exchange ideas freely. No one is elected or appointed to see that the conversation takes place in an orderly manner, and dialogue is seldom planned or scheduled.

These conversations usually take place wherever people happen to meet. The atmosphere is informal, and the words along with the manner of expressing them are conversational in mood and tone. The language you use in this atmosphere is comfortable, but perhaps not exact. Less precise words may be good enough because you use your voice, facial expressions, and gestures to make their meanings clear.

Plotting a Conversation

Some business conversations seem to be spontaneous and aimless. People get together and talk as if they had no real purpose in mind. This appearance may be deceptive, however; business or technical conversations usually have some aim, even if just to impress upon people the image we want them to have of us.

It is very helpful in starting a conversation to know what your main idea is. If you can, also decide whether you intend to inform or to persuade. If your purpose is to inform, you may want to give information or to get some; if to persuade, you may want to change your listeners' convictions or to induce them to do something.

Along with the main idea and kind of response you want, you should know as early as possible the kind of relationship you will need with your listeners to achieve your goals. To do this, it helps to know something about their background and interests. The more you know about these, the better you will be able to keep them interested, and the less likely you will be to say something that antagonizes them or causes them to "turn you off." Also, you will be able to establish the kind of mutual relationship, the common ground, essential for any worthwhile communication.

Holding Your Listeners

Here are some tips to help you be a better speaker in a business or technical conversation.

When conducting conversations about technical matters, first put your listeners at ease. Help them adopt a friendly attitude toward you and your ideas. One way to do this is to begin cheerfully. Say something pleasant about the listeners or about something they respect or admire, even if only the hometown baseball team. Even a casual remark about the weather may help develop the right kind of relationship.

Another way to make others in the conversation receptive to you and your ideas is to develop a warm, intimate relationship with them. Use their names as soon as you can. A person's name is a sweet sound, alerting him or her to what you may have to say. The musicians in an orchestra perk up and listen intently when a conductor raps a baton. A skillful conversationalist achieves somewhat the same effect by using names, first names when appropriate.

Be sure to keep the right relationship and listeners' interest throughout the conversation. Their willingness to pay attention may diminish shortly after the conversation starts. They often become distracted by their own concerns, or they may be disturbed by your peculiar mannerisms or expressions. Sometimes even your best friends won't tell you how they lose interest in what you are saying because of disturbing expressions such as "you know" or "ah" interjected after every few words. Be sure that your "body language" agrees with your ideas. Especially if you say something complimentary, be sure that your facial expressions and gestures do not distort or contradict your remarks.

Enable the other people in the conversation to feel they are important participants. Establish and maintain a comfortable eye-to-eye contact. Even when there is reluctance to look at you, keep trying to establish a sincere, friendly relationship by means of eye-to-eye contact. When several people are involved, you may have to look around at others so that no one feels ignored.

Consistently recognize your listeners as essential parts of a conversation. Allow them to feel that the worthwhile ideas resulting from the conversation are partially theirs. Periodically ask them by name to tell what they think about the ideas being discussed. Doing this will cause each to feel proud that you value what he or she thinks. Ask different ones questions that allow them to display their knowledge, to reinforce their self-esteem.

EFFECTIVE LISTENING

Most of the communicating we do about our work is done by listening, not by talking or writing. We spend only about 32 percent of our communication time talking about our work. We spend about 15 percent of it reading and only 12 percent writing. Consequently,

Effective listening

it is fair to say that effective listening, which occupies about 41 percent of our communication time, is at least as important as speaking. To succeed in business or industry, a person can use as much training in learning to listen as in learning to read and write.

Experts have pointed out that 75 percent of what a person hears is not completely or accurately understood. In other words, we understand only about 25 percent of what a person speaking to us intends. What appears to be our lack of intelligence on the job or elsewhere may be the result of failure to listen thoughtfully.

How to Listen Actively

You will begin to speak better when you learn to listen better. What you say to listeners is to some degree a response to what you detect from their reactions. Even when you are involved in a telephone conversation, you try to visualize the facial expressions and the gestures that influence your understanding of what the other person is saying.

Skillful listening may depend upon your ability to interpret silence. The intentional silence of the speaker sometimes says much more than words. Deliberate pauses between words change ideas and express different meanings. Good listening requires that you learn to accurately interpret what is implied by silence.

To be a good listener, learn to listen with your eyes and face along with your ears. Don't let your appearance give the impression that you are not especially interested in what the other people are saying. Once speakers detect your disinterest, they become irritated; you are threatening their self-respect.

As soon as you can, identify the speakers' purposes. Immediately try to determine whether each intends to inform or to persuade about something. Then point out for yourself the specific topics, the main ideas, and how they are related to the overall purpose of the conversation.

Be a good listener by listening actively, not passively. Good listening requires action. Don't just sit back to hear the words and watch the gestures. Do things, act, while listening. Communicate with the speakers. Tell them with your eyes and face the degree to which you understand. Silently question everything you hear. Is it a fact, a logical inference, or a reliable opinion, and how does the speaker intend it to be understood? Is the speaker sure about the reliability and qualifications of his or her sources?

Remember that a speaker can't talk as fast as you, the listener, can hear or think. Therefore, constantly adjust your listening pace to that of the speaker. A person speaks at about 150 to 180 words per minute. A listener's mind travels at about 500 to 1,000 words per minute. Consequently, there is much slack time during which a listener might be distracted. To keep your mind from wandering, periodically put it to work reviewing and summarizing what the speaker said. The more complex the speaker's statements are, the more helpful this reviewing will be.

It is much easier to listen to people when they are trying to inform us, to give us worthwhile and interesting information we do not already have, than when they want to persuade. When we try to listen to another person's argument in support of a conclusion, we often hear just what we want to hear, not what was actually said. To overcome this problem, keep your personal biases and feelings from coloring the speaker's statements. If, for example, a speaker makes a statement about "public office holders," don't translate it to "crooked politicians."

How to Disagree

Before you disagree with the speaker, be sure you understand what was actually said. If necessary, ask questions for clarification. When you are sure the speaker has made an error, ask leading questions to be certain that the error was made intentionally. You can restate exactly what you think you heard and then ask if that is what was intended. Doing this will enable the speaker to correct unintended mistakes.

Try not to challenge in an angry manner the accuracy of a person's statements. This will only make him or her defensively antagonistic. Avoid pointing out in front of others serious errors that will be embarrassing. If you can, call attention to serious mistakes privately.

When people in a business conversation become angry, let them talk themselves out before reacting. You might even want to encourage the venting of steam by sympathizing with them. To do this, put yourself in the agitated person's shoes so that you can understand why he or she is upset.

Begin working right now to overcome your bad listening habits. Once you have begun a conscientious program, you will be able to make more advanced strides toward becoming a truly effective conversationalist. Employees like supervisors "you can talk to." By that they mean those who also know how to listen.

GROUP DISCUSSIONS

A group discussion is a collective effort guided by a leader toward achieving predetermined objectives of mutual interest. Bringing several people together in a group discussion to contribute what they know about something creates a more forceful apparatus than a business conversation. The minds of two people are usually less effective than those of ten or twelve trying to solve the same problem.

Some of the same principles of human dynamics that apply in conversations are useful also in group discussions. Occasionally, however, they are somewhat more complex because more people are involved. Their efforts have to be organized and directed at a single objective in an orderly manner. Consequently, a group discussion requires advance preparation and planning to achieve its goals. The leader must be able to focus the minds of several people on a single purpose—to "spotlight" a more powerful intellectual force, a collective mind, on topics and problems.

Purposes of Meetings

Meetings are common forms of group discussions. There are several kinds, each organized in a manner suited to its intended purpose. The two main purposes are (1) informing and (2) problem solving. A small sales meeting or a large union meeting, perhaps a conference or convention, may be arranged either to distribute information or to arrive at solutions for problems the members have in common.

Many business meetings, large and small, are devoted to sharing information. This information may be about new products or new policies, laws, or other changes needing identification and explanation. The direct contact in a meeting is very helpful in achieving immediate clarification for a large number of people. Questions may be raised and answered so that much confusion can be avoided.

If complex issues, policies, and regulations are a part of the information to be given at a meeting, they should be distributed in advance in printed or typed form. This gives everyone a chance to study the information carefully, digest it, and do any research needed to make worthwhile contributions.

Problem-solving meetings are especially effective because they focus a large accumulation of knowledge and experience. The chairperson can help by first making sure there is a problem, identifying its causes, and suggesting tentative solutions. If someone at the meeting has already tested the tentative solutions and found one that works, that important information can be given quickly to everyone present. However, often the possible solutions have to be tested after the meeting and the proved ones distributed at a later meeting.

Problem solving is a more complex purpose for a meeting than informing. Informing only requires that information be given out and explained. Problem solving necessitates actually going through the steps of trying to solve a problem during the meeting. If the problem is solved before the meeting begins, what takes place during the meeting is informing—giving out and explaining the solution—not solving the problem.

Regardless of whether the problem solving takes place in a meeting or outside of it, certain things must be done to insure that the best and most practical solutions are found. The person leading, especially, should know what must be done in a problem-solving meeting to keep the collective mental efforts on the right track. Here is a summary of the important points:

1. *Immediately identify and define the exact problem.* At the beginning of the meeting, the problem to be solved should be identified precisely, including the reasons why it is a problem.

2. *Use audiovisual aids for clearer understanding of the problem and its dimensions.* In addition to movies, photographs, and recordings, use enlarged graphs, charts, tables, drawings, blackboards, etc.
3. *Clearly identify and explain all of the possible solutions.* Tell how they were found or otherwise arrived at. Each should be discussed in relation to those who have used it and the success they have had with it.
4. *Discuss how these solutions should be tested and judged for their effectiveness in solving this particular problem.* Indicate who is to do the testing and when the results will be known.
5. *Give the results of the tests at a later informational meeting.* Present the specific solution selected. Also, explain why each of the others was not suitable and give the evidence supporting the one chosen.

Types of Meetings

Each person involved in group discussions, whether they be informative or problem solving, should know something about the different types of meetings. This knowledge will help a chairperson select the best type of meeting for a given purpose. It will also help the participants know more precisely what they are supposed to do. Some types of meetings are better suited for informative purposes and others for problem solving. Some are organized to be equally suitable for either purpose.

Study groups usually are gatherings of several people to exchange information about something of common interest. The discussion is carried on in a free, open manner with a minimum of regulations. The leader merely greets the people, opens the meetings, and closes them, sometimes by summarizing the main points discussed. The main work of discussion is carried on spontaneously on equal terms by everyone present.

Many kinds of study groups are used in business. The men and women in a department store may meet periodically to exchange ideas about improving their sales techniques. Even a large conference may be broken up into small study groups to discuss new products on the market.

Workshops are just another kind of study group often used to distribute and explain information. Since they are small groups, a warm, friendly relationship develops among the participants, leading to a freer discussion. Workshops do not meet regularly; they are called when there is a need for them. Often they are conducted as a part of a larger meeting, perhaps a conference or convention.

Round table discussions are small group meetings that are used either for informative or problem-solving purposes. The participants talk "around a table" (not necessarily a round table) so that all, especially the more timid ones, are more obligated to contribute what they know. As the discussion goes around the table, each person is expected to add something that has not already been presented. Those who speak last carry a heavier burden because they are expected to contribute something that hasn't already been said.

Committee meetings often are small group meetings. The people who attend usually represent a subgroup of a larger organization. All of the committee members belong to the same larger group. Usually a committee is assigned a certain subject to discuss and then to report on it, along with any recommendations, to the larger organization. Consequently, the committee doesn't make the important decisions and value judgments about problems; it is up to the larger group to make them.

Conferences are large or small gatherings conducted in private to distribute needed information or to arrive at important decisions. The persons at a conference represent two or more different departments within a company or within a different organization concerned about the same things.

Conventions are merely large conferences. They are held to apply collective knowledge, intelligence, and experience, to make decisions, and to exchange information. They are also used to make or renew acquaintances. Sometimes they are convened just so people can compare notes about a situation with others concerned about the same thing.

CHAIRING A MEETING

A person's ability to lead others in a meeting is just as important, and may be more important, than the amount of technical knowledge and education he or she has. One person with technical knowledge or skill equals only one in "people power," if we may call it that. However, that same person effectively communicating that skill and knowledge to a group of twenty others in a meeting multiplies in effectiveness into twenty-one "people power." It certainly is worthwhile for an employer to get twenty skillful employees for the price of one.

The ability to lead a meeting places a chairperson under management's spotlight. A meeting is a good place for a person to demonstrate ability to lead and motivate a group toward achieving the organization's goals. Chairing a meeting gives a man or woman a chance to display communication skills in managing others to those above him.

All meetings on company time are expensive when they take people in the higher wage brackets away from their regular duties. Successful meetings are expensive, but unsuccessful ones are much more expensive because they are wasteful. Poor leadership by the chairperson is often the cause of unsuccessful meetings. If the person in charge doesn't know how to guide or motivate, some people in a meeting will find it difficult to overcome the apathy or confusion that may result. The meeting will not be a collective effort successfully directed at achieving the meeting's objectives.

The Chairperson's Tasks

The most difficult part of a chairperson's task is keeping the meeting discussions moving smoothly. The confidence and cooperative atmosphere established with the rapping of the gavel at the beginning should be maintained constantly. Throughout the meeting, the leader must maintain a free, cooperative exchange of related ideas by always being ready to handle personality conflicts, quarrels, and unprofitable arguments.

A chairperson must be especially careful that the meeting does not become a debate forum. A person addicted to debate regards almost everyone as an adversary, as an opponent upon whom he or she must pounce for making the slightest error. For that type of person, a meeting is a form of competitive sport, an activity in which people diligently try to outwit each other.

Wholesome competition is required by most human activities, but cooperation is more effective in achieving communication. A business meeting should be a cooperative effort. When someone makes an unintentional mistake during a meeting, the leader should see that enough time is allowed for that person to correct the error. The others should not be allowed to interrupt and attack. The chairperson or another participant should help the person making the error to recognize the mistake—perhaps by asking leading questions—and to correct it. This same kind of help should be given to anyone who has introduced irrelevant ideas into the discussion.

To help avoid unprofitable disagreements, the chairperson might ask every person challenging or disagreeing with anyone to begin by restating the other person's statement, so as to clearly identify the idea not acceptable. This approach will prevent misunderstandings and will encourage cooperative effort.

Sometimes the agenda scheduled for a certain meeting may not be completed as planned. When this happens, the leader must see to it that someone takes notes identifying all the unfinished business to be completed at a later meeting.

The leader must be sure also to make arrangements to follow up the decisions made, including the motions voted on and passed. This is done to be certain what was decided is carried out. If, for example, a decision was made that the people not attending the meeting be informed of duties assigned them, then the chair must see that someone does that. Since the pressures of business may cause a busy chairperson to forget, the leader must learn to give some of the follow-up duties to others.

The leader of a meeting should be aware of any smoldering antagonisms caused by differences among members of a meeting or of a dominant clique. An aggressive participant may be outspoken, while timid ones may be sensitive and easily offended. The leader should always be on the lookout for those people who intentionally or unintentionally intimidate others. This is often the most damaging trait of some very worthwhile but domineering people. Unless they are handled properly, they may discourage others from making worthwhile contributions. The bully may anger some, who then may cause nothing but confusion by antagonistically expressing how they feel.

Preparing and Opening a Meeting

As a chairperson, you will find that making the proper preparations will insure a more productive meeting. Plan the meeting carefully. Predetermine a convenient time and place. Be sure the atmosphere and furnishings will

Chairing a meeting

stimulate cooperative effort. Also, make arrangements to see that any needed equipment, such as projectors, screens, and tape recorders, will be ready for use when the meeting starts.

Prepare an agenda and, several days before the meeting, distribute a copy to the other participants. It should contain the time and place, both to notify the unaware and to remind anyone who may have forgotten. The agenda should clearly tell what the specific purpose of the meeting is: to inform, or to solve a certain problem. List specifically the items to be discussed. Sending the agenda in advance of the meeting should enable the participants to do any needed research. Consequently, they will come better able to contribute toward the meeting's purpose.

While the participants are arriving for the meeting, "psych" them into wanting to contribute to the joint effort. Be sure that a friendly, cooperative atmosphere exists. Move about, perhaps while coffee is being served, greeting as many people as you can.

Open the meeting decisively and purposefully. Welcome those present and thank them for attending. Be sure everyone has an agenda, restate the purpose of the meeting, and ask if there are any questions about the agenda, including the sequence in which the items appear on it. Ask if anyone wants the sequence changed. If the change is agreeable to the others, make it.

Start the discussion of the agenda items by explaining the importance of the first one. Provide any essential explanations: definitions, case history, and other background information. Do as much of the same kind of thing for each item as you think necessary.

Guidelines for Managing a Meeting

Listed below are a few of the more important principles that will enable you to conduct a successful business meeting.

1. *Keep the meeting on the right track.* Do not allow lengthy detours from the issue being discussed or from the predetermined purpose. When unrelated or only slightly related ideas are brought in, gently guide the discussion back to its proper focus.
2. *Do not dominate the meeting with your own ideas or personality.* Don't impose your ideas over those of others. Instead, present your suggestions in question form to arouse member responses. Use these questions to motivate the silent persons to become active, to make some contributions. Make them feel their importance, your respect and need for their answers.
3. *Do not allow anyone to intimidate or bully the others.* Avoid confronting the intimidator during the meeting. Do it later during a private conference. During the meeting, divert attention by saying something humorous or distracting or by asking someone else to make a comment. Be especially watchful for rude implications, offensive remarks, emotional quarrels, and personality conflicts.
4. *The participants should be allowed to make complete comments without interruption.* A good policy to establish is that every

contributor be allowed a second or two after his comment for an afterthought, something that should be added to clarify or correct the statement. This will save time wasted in controversy.

5. *Do not condemn a person who makes an error or allow anyone else to do it.* Help out by asking a leading question, enabling the person to see and correct the mistake. Also, you might restate exactly what was said and ask the person if that is what was intended.

PARTICIPATING IN A MEETING

The people taking part in a meeting are just as essential for its success as the leader. A chairperson can't do the job alone. The leader needs the knowledge and cooperation of the others if the meeting is to achieve its objectives.

Just as a meeting serves as an ideal arena within which a chairperson can display ability to manage a group of people, it also gives the other members a chance to demonstrate their skill and willingness to work cooperatively. It lets them show the leader, who often is a representative of management, and other supervisors their potential for promotion. The participants in a meeting reveal the extent and depth of their technical knowledge along with how well they work with others. They also indicate their attitudes toward the organization by their willingness to serve its interests.

The leadership suggestions previously listed for the chairperson should also be studied by everyone else. Knowing what the chairperson is trying to do enables the others to help. To have a successful meeting, the work of the leader and of the participants must mesh.

Listed below are some ways you can be a good participant in a meeting.

1. *Study the agenda and do any needed research before the meeting.* This will enable you to make worthwhile contributions.
2. *Get to the meeting before it starts.* This will enable you to take your seat without coming in late and interrupting the meeting. It also will give you time to review the agenda and any notes you may have.
3. *Present your comments calmly, courteously, and cooperatively.* Avoid participating in quarrels. Do what you can to help the leader resolve them. Try not to use sarcasm or name-calling to bully others.
4. *Express your ideas briefly and clearly.* When under the spotlight, a person sometimes gets nervous and finds it hard to stop talking.
5. *Avoid interrupting.* Let the person making a comment express it without interruption and without pressure. When a speaker sees gestures and facial expressions intended to rush him or her, anger increases and cooperation decreases.
6. *Be recognized before speaking.* At least indicate to the chairperson that you want to say something by giving some kind of signal.
7. *Try not to introduce irrelevant ideas.* When others do, help the leader by doing what you can to get the discussion back on course.

Participating in a meeting

8. *Show respect for the ideas of the other participants.* There never is need for shouting, abrasive language, rude gestures, or other offensive behavior.

Unless the leader of a meeting understands human nature and relies on the ideas discussed on the preceding pages, little success will be achieved. A bit of study and practice in using these ideas and those that follow will be thoroughly worthwhile. Many of these same principles are needed for success in all organized human activities, not just business meetings.

When you act out some of the following applications, your instructor may decide to tape them. Closed-circuit TV, if available, will allow students in another room to evaluate your demonstrations. Be sure to prepare them carefully.

Application 3–1

Select one of the situations listed below and write a dialogue you can present in front of the class with the help of one or more of the other students. Your instructor may assign some of the class members to evaluate the person starting the conversation. Other students may be assigned to evaluate the participants in the conversation who are mainly listening.

Your instructor may choose to evaluate the dialogue demonstration by having the whole class discuss it after its presentation. You may also be asked to write an evaluation of the dialogue, its effectiveness, and the skill of the students involved in it. Therefore, it would be a good idea to take notes while each demonstration is being presented.

Whether you are the person responsible for presenting the demonstration or an evaluator, be sure to tell the kind of business the conversation is about, the positions of the participants, where the dialogue takes place, and the purpose of the exchange.

1. A vacuum cleaner salesperson is trying to sell a new model to someone who does the cleaning in a residence or business.
2. A customer is trying to return a damaged or defective suitcase to a clerk in a large department store's luggage department. The customer should be satisfied when leaving.
3. A police officer is giving a traffic ticket to a driver who reacts angrily. The driver should become more understanding of the officer's fairness at the end of the dialogue.
4. An auto insurance claim adjuster is trying to convince a claimant that the company settlement offer is fair, although it won't cover the full cost of the repairs. Try to find sound reasons to satisfy the claimant that the insurance company is doing what it has to do to protect all of its policyholders.
5. One friend succeeds in persuading another to lend a new car for a date or special occasion.
6. The manager of a small restaurant or cafeteria is trying to persuade an employee to substitute for another on Sunday or on a holiday.

Application 3–2 Review the pages in this chapter dealing with impromptu conversations and be prepared to participate in one before the class. Your instructor may tell you to participate in a dialogue between a gas station owner and an employee, a nurse and a patient, etc. Your instructor may choose to tape your conversation for later class evaluation.

Application 3–3 Prepare an outline or a list of things to say to a class to explain how they should do something. Present your talk as briefly as you can, but clearly. Ask the class to prepare a summary of your main points in a written outline or list. Then write your own on the board so that they can compare theirs with yours to determine how well they listened and how well you expressed your ideas. Also, ask the class to evaluate the way you delivered your presentation, especially to point out the things discussed in this chapter. Your instructor may choose to collect these outlines to check how well you listen.

Application 3–4 Read a brief report or letter to the class, then ask the students to:

1. Give the name and address of the writer or the name and address of the receiver, and the date.
2. Identify the purpose of the report or letter.
3. Give the main points of the report.
4. Briefly describe the way the report is organized.

Do you think that your audience did a good job of listening? Also, ask them to give you feedback on how well you orally expressed what you had to say.

Application 3–5 Make a table or pie chart showing the approximate amount of time spent communicating on a certain day either at school or on the job. Let the table or chart show the amount of time (a) speaking, (b) listening, (c) writing, (d) reading.

Application 3–6 Observe closely the kinds of oral communication that take place at work. (If you are not employed, spend some time observing at a local bank, grocery, drugstore, gas station, or some other place.) Try to take notes accurately, identifying the characteristics of the communication and its basic type (informative or problem solving). Now write an evaluation pointing out what you think is good or bad according to the suggestions in this chapter.

Application 3–7 With the aid of three or four students from the class, prepare a meeting agenda for one of the situations listed below. Distribute copies to the entire class and conduct the meeting as a chairperson. After calling the meeting to order, explain its purpose and provide any needed background information. Have the class members not in the demonstration take notes so they can evaluate the way the meeting is conducted, using the principles discussed in this chapter as criteria.

1. Sales in your department of a large store have declined. Conduct a meeting to identify what is causing the decline and get suggestions about how to increase sales. Besides yourself, at least five other persons should attend. Assign someone to take notes and to read them back when the

meeting is finished. This should be an informal meeting, but conduct it in an orderly manner.

2. A grocery store manager has several clerks responsible for marking prices on the packages as they stack them. One of them is consistently slipping up on the job. You are sure you know who it is, but you don't want to offend and perhaps lose that person as an employee; consequently, you avoid confronting directly but call a meeting instead. Using several students from the class, conduct an informal meeting for this purpose. Ask the remainder of the class to evaluate the participants and to take notes so that they can make suggestions to improve the participation.

Application 3–8

Attend a group communication situation, a meeting, conference, class seminar, or an ordinary class. Observe and take notes. Write a brief evaluation of how the principles studied in this chapter contributed to the success or failure of the group discussion. Also, pay attention to the quality of the listening done and explain whether it was due to the conversational ability of the chairperson and other speakers or to the listening skills of the other participants.

Application 3–9

You are the manager of a large grocery or other business. Recently you were told by the main office that your branch will have to remain open twenty-four hours a day. You will have to hold a meeting to decide which of your employees will work at night. List the steps you will be prepared to take to overcome resistance and to induce some employees to volunteer. Prepare a notice of the meeting and an agenda to be distributed to everyone in the classroom.

Basic forms: enumeration and analysis

4

the main divisions of a communication . . . their functions . . . the basic forms of development . . . how they differ . . . kinds of relationships they develop: structural, functional, causal . . . how to use them

A writer or speaker usually follows certain patterns of development in organizing and expressing ideas. Knowing what these patterns are and what each is supposed to do will enable you to speak and write clearly.

MAIN DIVISIONS OF A COMMUNICATION

Most written or oral communications have three main divisions. Each performs an important function. These divisions and their functions are outlined below.

A. *The Beginning Section*

1. Establishes contact with the intended receiver by arousing a receptive, cooperative relationship.
2. Identifies what the communication is about and what the writer or speaker mainly intends to say.

Main divisions of a communication

B. *The Developing and Supporting Section*

1. Adapts the main ideas of the beginning section to the intended reader's or listener's background, enabling her or him to understand them.
2. Provides enough reliable information when the main purpose is to inform.
3. Provides enough supporting evidence from qualified sources when the purpose is to persuade.
4. Amplifies and develops the main ideas of the communication.

C. *The Concluding Section*

1. Summarizes the main ideas.
2. Restates important findings, conclusions, and recommendations.
3. Maintains and reinforces the desired relationship.

You know how to establish a receptive attitude when meeting acquaintances. You know how important it is to greet them in a pleasant manner. Usually the greeting is followed by a comment about the weather or something else of mutual interest. In technical communication, you get to the "something of mutual interest" immediately. That might be some worthwhile information about a business or industry in which the receiver is involved, or it might be the solution for a problem. This opening may be followed immediately by background information defining the main idea or stating how the receiver will benefit from what the communication says.

The concluding section summarizes and restates the important ideas you want the reader to be sure to remember. When your purpose is to inform, it will contain in brief form the main points of information, your important findings. When your purpose is to persuade, your summary will contain your main reasons and evidence along with important conclusions and recommendations.

FORMS OF DEVELOPMENT

Our concern here is the development section of a communication. To achieve your intended purpose, you should know how to expand and clarify the main ideas you identified in your beginning section. You should know as much as you can about the important forms or patterns in which your supporting ideas are developed to achieve your goals. We have classified these forms into three kinds: basic, general, and supporting. In this chapter we will focus attention on the basic forms of development. Chapters 5 and 6 will discuss the general and the supporting forms.

Basic forms are patterns of organization by which you will develop your general and supporting forms. The two main ones are enumeration and analysis. Regardless of which general form you decide to use in developing your ideas, you will have to start out with enumeration or analysis. When you want to tell what your subject looks like, tell how it developed into what it is, or tell the reasons for its appearance or behavior, you will start out by deciding whether to enumerate or analyze.

ENUMERATION

Enumeration, unlike analysis, is merely laying down details about something next to each other without stating the relationship between them. You may list the components or ingredients in the subject without telling how they are structurally connected. You might give the series of things the subject does or the actions its components perform without explaining how they work together, their functional relationship. Or you may choose to enumerate the reasons that enable the subject or its components to do what it is able to do without giving the reasons why they look the way they do or why they are able to perform their functions. This will become clearer to you as you read the examples and then the explanations of structural, functional, and causal relationships later in this chapter.

Enumeration may be developed by inserting a series of single words or phrases within one or more sentences. Notice how this occurs in the second sentence in the following supporting paragraph:

> Medical research now links sugar with several serious human illnesses. Hypertension, diabetes mellitus, and a variety of heart and kidney diseases have been traced back to overconsumption of sugar. "Sugar is a slow poison to millions of people," wrote William Duffy, author of *Sugar Blues.* The average American consumes over a hundred pounds of sugar per year.

The following paragraph expands the enumeration in the preceding illustration into a series of similarly constructed sentences. The intentional repetition of the word "sugar" also indicates the deliberate development of enumeration.

> Medical research now links sugar with several serious human illnesses. A close relationship has been discovered between the overconsumption of sugar and hypertension. Medical evidence indicates that excessive use of sugar triggers early diabetes mellitus. A correlation has been found between the overuse of sugar and a variety of heart ailments. Some authoritative medical literature clearly suggests that excessive consumption of sugar may be a contributing factor in some kidney diseases. "Sugar is a slow poison to millions of people," wrote William Duffy, author of *Sugar Blues.* The average American consumes over a hundred pounds of sugar per year.

Here is another example. Notice how the writer uses a series of words and phrases in several of the sentences:

> Our blender is capable of several different functions. The usual purpose of a blender is to mix different liquids. A blender is especially useful in making beverages, from a Mocha Frappe to a Tom Collins. The blender also chops meat and vegetables effectively. Hamburger, shrimp, corned beef, ham, and so on, can be thoroughly diced and chopped by whirling blades, saving much time and trouble. Similarly, carrots, cabbage, lettuce, radishes, etc. may be quickly and easily combined into many kinds of sauces and salads. The blender may also be used for mashing food for babies, grinding coffee beans, and crumbling bread, cookies, or crackers for use in baking.

Below is another illustration of the use of sentence enumeration in supporting paragraph development. The similarity of the sentence structure, as in the other example, clearly indicates the writer's deliberate use of enumeration. The words "should be" are used in almost every sentence to tell the reader that the writer is enumerating intentionally.

> Industrial management can help to prevent injuries to welders by informing them of the potential hazards and how to avoid them. Posters should be placed in strategic locations to inform about safety and remind workers about safety measures. Local or general exhaust ventilation should be provided to control air contaminants as required by the various types of welding processes. Personal protective equipment, including eye goggles, helmets, and hand shields, should be provided. Welding operations, when possible, should be isolated from other industrial activities, particularly from degreasing tanks and solvent cleaning operations. Even with the level of trichlorethylene well below the maximum safe limit, there is the potential for dangerous concentrations of phosgene or other vapors to occur. Good housekeeping practices should be maintained throughout welding work areas.

To focus more emphasis on the enumerated items, you may place a letter or a number in parentheses before each of them for emphasis. Here is an example:

> Whenever a car can't be started, three things should be checked first. (1) See if there is gas in the tank. (2) Turn on your lights to see if your battery has enough power, (3) Trace all ignition wires for bare or broken ones and for loose connections. If you still can't discover the cause, call a serviceman.

Another way to gain still more emphasis is to set off and indent the enumerated items in a numbered list. For example:

Whenever a car can't be started, three things should be checked first:

1. See if there is gas in the tank.
2. Turn on your lights to see if your battery has enough power.
3. Trace all ignition wires for bare ones and for loose connections.

If you still can't discover the cause, call a serviceman.

All of our communications result from thinking, and thinking is the mental process of establishing relationships between ideas. Three of the most common kinds of relationships between ideas are (a) structural, (b) functional, (c) causal.

The main difference between enumeration and analysis is in the way they indicate the structural, functional, and causal relationships between their details. In enumeration, these relationships are just implied by the context, but in analysis the writer clearly explains these relationships. This difference will become much clearer after you have studied the following sections and examined the accompanying illustrations.

Structural Relationships

Structural relationships in enumeration are expressed by listing the parts, material, or ingredients of something without carefully explaining the interrelationship between them. This kind of enumeration is often used in description. Here is an example.

> The Boeing 747 is one of the largest and most comfortable jets built today. It is 73 yards long, 65 yards in wingspan, 21 yards in height, and weighs 710,000 pounds. This plane is equipped with four Pratt and Whitney JT9D engines, with each engine having 43,500 pounds of thrust. In the coach section, passengers sit nine abreast. Two aisles go down the length of the plane, dividing each row of seats into one two-abreast, one four-abreast, and one three-abreast. There is a separate lounge in the upper section. Pilots claim that the plane is more comfortable than smaller jets and that it handles more easily.

An organization such as a team, family, corporation, or army has no structural components. Its parts consist of people (employees, relatives, etc.); therefore, organizational enumeration is a kind of structural enumeration.

Functional Relationships

Functional relationships in enumeration explain how the subject or its parts work, what they do. Everything that functions, exists, or happens does so in a time sequence (first, it does this: next, that happens, etc.). Functional enumeration may be just listing the functions or uses of the components of something or telling the chronological order in which something is done or happens. Following is an example of this kind of enumeration.

Enumeration

A variety of devices are installed in today's automobiles to make driving more comfortable. Air conditioners are installed in them to keep a person cool in the intense heat of the summer. Stereo tape players are furnished to provide the kind of music enjoyed by the passengers. The seats are made with deep foam rubber for comfortable riding. Heavy duty shock absorbers and springs are on all deluxe models to reduce the bumps when driving over rough roads. Even body and engine sounds are minimized by effective soundproofing. All of these additional comforts can be had in any car for a reasonable charge.

Following is an example of a functional enumeration telling how something is done in an orderly time sequence:

Changing the oil in an average American car can be done with few tools. First, place a jack under the front of the car, raising it just high enough to crawl under. Place jack stands under the front axles. Put one of your old pails directly under the oil pan and remove the drain plug. This plug requires a ⅞ or $^{13}/_{16}$ wrench. Next, you must locate the oil filter. Usually, it is a white, red, or blue cylinder 4×6 inches protruding from either side of the engine block. To remove this filter, you should turn it counterclockwise. After replacing the old filter with a new one, replace the drain plug. Next, remove the jack stands, release the jack, and lower the car. Finally, be sure to replace the oil drained from the engine with five or six quarts of clean oil.

Causal Relationships

Causal relationships in enumeration are a series of causes or reasons explaining why something exists or happens. They may be a series of theories as to the causes for the origin of the subject. Following is a causal enumeration offering several different causes for the origin of life.

Scientists disagree regarding the origin of life on earth. Some say that life is a natural state of matter and, therefore, was impossible to prevent; others say that its emergence was largely by accident or according to the laws of probability. Scientists disagree also about the location where life could have arisen. Some believe it was from dissolved molecules in the primordial soup of the oceans or in shallow pools. Others believe that discrete systems were formed by absorption of particles of minerals or clay which served as catalysts to activate polymerization and biochemical reactions characteristic of life. Another theory says that during floods organic molecules could have lodged in cracks and crevices, and subsequent rigors of wetting and drying cycles would have increased the chances of polymerization and mutation.[1]

Enumeration is an important pattern of development because it will allow you to deal with a broad range of time or space in fewer words than

---

[1] Eugenia Keller, "The Origin of Life," *Chemistry*, December 1968, p. 10.

analysis. It will enable you to skim over topics by merely listing the details related to your subject. It would be difficult for you to explain the rock formation of all the North American Continent in a report of a few pages without enumeration. The following three-paragraph section (a section is a developing form consisting of more than one paragraph) illustrates how important enumeration is in dealing with a broad range of time.

> As improbable a candidate for automotive immortality as ever trundled down the pike, the Tin Lizzie resembled a do-it-yourself kit put together by somebody who couldn't follow instructions and had lost most of the pieces anyway. It lacked just about everything considered essential on a modern car, from bumpers to spare tire. There was only the stamped outline of a left front door on the early ones. The gas tank was under the front seat, and to check the oil, you had to crawl below and fiddle with a pair of petcocks on the rear of the crankcase. But it sold. It sold like hotcakes, and if it didn't make Henry a billionaire, it brought him mighty close.
>
> Ten years after its introduction, half the cars on American roads were Model T's, a ratio which appeared in many guises: delivery vans, garbage trucks, buses; with flanged wheels railroads used them as inspection cars. During WWI they served as staff cars and ambulances. Hollywood made them collapsible and featured them in slapstick epics.
>
> It was in the rural areas, however, that the Model T revealed its unique versatility. A farmer buying one of Ford's flivvers drove home with a lot more than mere transportation. He used it to pull a plow in the spring and harvester in the fall. By jacking up the rear wheels and attaching a pulley, he could pump water, saw wood, grind feed, shell corn, shear sheep, or produce electric current. When the chores were done, he would load the family in front, produce in back, and would go off to market with several crates of chickens lashed to the running board.[2]

You may have to rely on enumeration when informing people who are not experts in a field. When you communicate with them, you may skim over the surface by discussing lightly only the important points, without penetrating any of them deeply. By doing this, you will give the readers what they need to understand, what you mainly want to say. Doing it that way will enable you to avoid confusing anyone with too many complex details. If, for example, you want to inform people about the way synthetic rubber is manufactured, you may have to skim over the chemistry involved to give enough background information without confusing the reader with complex chemical analysis.

Application 4–1

All of the following are intended for development by enumeration. Read each and tell whether it calls for structural, functional, or causal enumeration.

---

[2] Excerpted from *The American Legion Magazine* (Copyright 1968) by permission.

Enumeration

1. A flashlight has only a few main parts.
2. Here are the main reasons why I want to be a mechanic.
3. Each morning, a nurse must do certain important duties.
4. Certain reasons are usually given for business failures.
5. Following are the main parts of a drawing table.
6. To roof a house, you need several important tools.
7. Eating a variety of vegetables is healthful.
8. The layout of the library is simple.
9. Here are five steps to follow in taking inventory.
10. Being courteous to tenants is profitable.

Application 4–2 Write two separate paragraphs about "wallets." Develop the first by structural enumeration and the second by functional enumeration. Be sure to indicate which is intended as structural and which as functional.

Application 4–3 Select one of the following and write a structural, a functional, and a causal enumeration for it. You may have to do a little research for each. Be sure to tell what predetermined purpose you intended for each. Also, tell which kind of enumeration each represents.

1. Any house appliance.
2. A suit of armor.
3. A carpet or rug.
4. A gun.
5. A windmill.
6. A piece of furniture for house or office.
7. A kind of artificial bait used in fishing.

Application 4–4 Write a structural, a functional, and a causal enumeration for the accompanying drawing of a handscrew clamp. Limit each to one paragraph. Tell which type of enumeration you intend each to be.

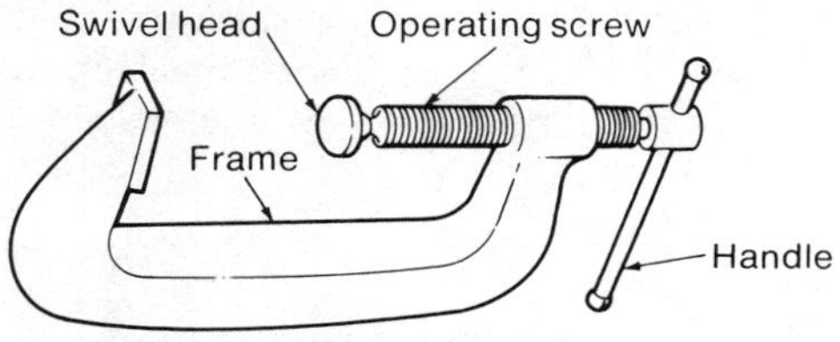

HANDSCREW CLAMP

ANALYSIS

Analysis is the taking apart of components, functions, or causes for something while explaining the relationships among them. As you learned before, enumeration does not clearly state the relationships among the items in the series. In analysis, you will want to point out explicitly the structural, functional, and causal relationships that enable you to treat the subject in depth, to penetrate it more.

Structural Analysis

Structural analysis not only identifies the parts of the subject; it also tells how the parts are related in space (above, below, before, behind, etc.) and/or how the parts are joined (cemented, nailed, welded, etc.). The word "parts" includes the structural ingredients or materials making up the subject discussed. An example of a supporting paragraph of structural analysis follows:

> A piano can create its pleasant variety of musical tones because of many intricately connected parts. When a key is pressed, its other end rises and pushes the hammer against strings set in a cast iron frame, to withstand tension. The striking hammers are attached to a device on the far end of the keys. This mechanism catches the hammers on the rebound and holds them poised for the next strike. A tone-reinforcing soundboard consisting of a thin piece of wood is attached to a frame under the strings.

Although the illustration above does contain a bit of explanation of the functional relationships, the main focus is on explaining the structural ones.

Following is one more example of structural analysis:

> An adjustable triangle is used in drafting to draw lines at certain angles. It is made up of two main parts, the body and adjustable side. The body making up two sides of the triangle is often made of clear plastic. The adjustable side is the third side of the triangle; it also may be made of clear plastic. Riveted to the adjustable side is the face plate, usually made of white plastic. The face plate has markings for every degree along with number markings for every ten degrees. This plate is attached to the body by means of a screw and shoulder nut. There is also a lock nut attached to a flat head screw which passes up from the bottom of the body through the face plate. A degree marker, made out of a small round piece of plastic with an arrow or line scribed on it, is riveted to the body.

Functional Analysis

Functional analysis divides, separates, dissects, or takes apart something in an orderly manner to explain mainly how it functions, works, or does what it is supposed to do. In enumeration, you just list the functions performed, but in analysis you should also explain the relationships between them.

Below is a supporting paragraph developed by functional analysis. Notice that although it does contain the names of the components, it goes beyond structural analysis; it also tells how the parts function in conjunction with each other.

> The Humming Bird is a sports car especially built for high speeds and tight turns. It has small wide tires, which hold the road in tight turns. The tires are built with steel cords running the length of the tires to increase their roadholding power. The axles have a tight differential, giving the Bird a very tight turning radius. The gears are best suited for high speeds and tight turns, as they have a low ratio for lower gears and a high ratio in higher gears. The drive shaft is connected to the light, powerful engine with the necessary strength in the universal joint needed for high-speed bumps. The engine is small, but powerful, as its small 90-cubic-inch displacement develops a very efficient 90 horsepower. The efficiency of a horsepower per cubic inch is necessary for tight turns at high speeds. The carburetor mounted on top of the engine is best suited for the small engine. It has four barrels, one for each cylinder, which feed an ample amount of high-octane gas to the carburetor. All these parts are combined into one unit—a light, efficient sports car that may be seen in the lead in many races.

Here are two very helpful illustrations of functional analysis. In the first one, dealing with the human respiratory system, the writer not only explains the job or function each part performs, but also the orderly time sequence in which it happens.

> The human respiratory system functions precisely and reliably without the conscious control of a human being. First the air is drawn into the body through the mouth and nasal passages by the action of muscles contracting and expanding under the lungs. After passing through the bronchial tubes, the oxygen is blown into air sacs which resemble clusters of grapes. The capillaries, tiny blood vessels connecting the arteries with the veins, pick up the oxygen and transport it to the main arteries to the heart, which pumps it to the body cells. While this is taking place, waste gas, carbon dioxide, is transported from the cells to the veins, through the heart, and then to the lungs to be exhaled.

The second, dealing with water purification membranes, also illustrates the closeness between the functional and chronological aspects in this kind of analysis.

> New water purification membranes, which are important in helping to solve the water pollution problem throughout the country, are made with two basic layers. The first consists of a cellulose acetate product forming an extremely dense layer. As polluted water passes through it, this cellulose layer removes the impurities from the water. The second layer is in direct contact with the first, but on the opposite side. The water that has been partially purified after running through the first layer then runs through this second one of more porous material and is purified further. These two layers work together through diffusion. One serves as a heavy purification layer and the other for more refined purification.

Causal Analysis

Causal analysis, like functional analysis, takes something apart in an orderly manner, but gives causes and reasons why it is designed or constructed that way, why it functions as it does, why it originated and exists or stopped existing. The writer often uses causal analysis to support conclusions and value judgments. Causal analysis may be used to explain the causes for a problem and why a certain solution for it is worth consideration.

When writing an effective causal analysis, you probably will do the following: (1) identify the problem or situation you are attempting to explain or solve, (2) state the factors, evidence, symptoms, or inferences that explain the causes of the problem, (3) tell how you determined that these causes are related to your conclusion. Here is a section of this kind of causal analysis.[3]

> *Problem identification*
>
> For many years scientists have been trying to explain how our universe came into being. . . . Dr. Joseph C. Hackman of Ivy University believes that he has a way of verifying the belief that our planets resulted from stars that have cooled off to form the solid matter of a planet.
>
> *Supporting causal evidence*
>
> Recently, Dr. Hackman and his assistants discovered high-energy gamma radiation pulsating at a steady 30 beats per second from a star "dying" deep in the Milky Way. The source of these gamma rays is a "pulsar," a tightly packed segment of a star which broke apart in A.D. 1045. The energy shooting out from this pulsar, located in the Crab Nebula, causes the pulsations detected by Dr. Hackman and the other scientists at Ivy University. They believe that the steady pulse of the ray is caused by the rotation of the pulsar. They also believe that the slowing down of these pulsations, a few millionths of a second a day, indicates that the star is dying, cooling off. Therefore, they believe that measuring the decreasing pulse rate more accurately should enable scientists to better explain how the planets in our universe, including the earth, were formed from "dying stars."
>
> *Conclusion*

Here is another example of causal (or critical) analysis, but this one deals with a less technical problem than the preceding example.

> After ten months of careful investigation, the building inspectors concluded that the collapse of the seventh floor of the ten-story office building was caused by structural defects. The ground upon which the building was erected had a deep pocket of soft sand. Because the design of the concrete supports under the building did not adequately spread the weight of the building, it rested on only a few of the pillars. This

---

[3] Reminder: a section is a form of development consisting of more than a single paragraph.

situation caused the concrete support pillars to sink into the soft sand, causing one side of the building to sink slightly. Although the structural steel framework under all of the floors, except the seventh, was able to withstand this shift in weight, the framework under the seventh floor was either not assembled properly or the structural steel itself was defective. Based upon these findings, the investigating committee could only conclude that the cause of the collapse must be attributed to these structural weaknesses resulting from inadequate land preparation.

ENUMERATION AND ANALYSIS COMPARED

The following paired supporting paragraphs illustrate the similarities and differences between enumeration and analysis. The subject is the Eddy bow kite. How it is constructed is described in the structural paragraphs, how it is flown in the functional paragraphs, and what makes it fly in the causal paragraphs. As you read, compare the different approaches.

Example of Structural Enumeration

An Eddy bow kite is a simply constructed device. It consists of two thin strips of wood, one about 24 inches long and the other about 32. These are fastened together, and a single piece of string is attached to the four stick ends. A diamond shaped sheet of paper is pasted over the kite frame. A strong piece of cord is stretched tightly from one end of the shorter crossmember to the other. A strip of cloth about three or four feet long and two or three inches wide is tied to the bottom end of the spine, the longer stick. The line for the kite is attached at the spot where the two sticks cross. Refer to the accompanying drawing.

Example of Structural Analysis

An Eddy bow kite is a carefully constructed device. It consists of two thin strips of wood, one about 24 inches long and the other about 32 (Figure 4.1A). The longer piece is placed vertically, and the shorter one is set horizontally in a notch cut about eight or nine inches from the top of the longer one (B). At this juncture, the two pieces are bound tightly with a piece of string (C). A continuous piece of string is stretched around this cross-shaped frame and inserted into the slits in the four ends of the sticks (D). These slits may be taped or tied with a short piece of string. Over the frame, a piece of diamond-shaped paper folded an inch or two around its border is pasted. A bow is made by a strong cord fastened to one end of the horizontal stick, stretched tightly across, and attached to its other end. A strip of cloth about three or four feet long and two or three inches wide is tied to the bottom end of the spine, the longer stick. The line for the kite is attached at the spot where the two sticks cross (E).

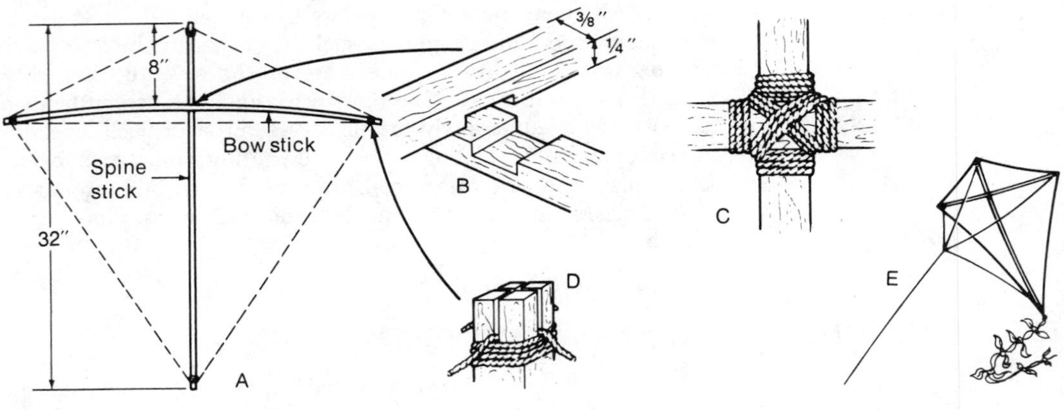

Figure 4.1
Eddy kite construction. (A) frame; (B) center notching; (C) center lashing; (D) stringing the perimeter; (E) attaching the string.

Example of Functional Enumeration

To fly an Eddy bow kite skillfully, a person should know something about the way it functions. Two thin strips of wood, one about 24 inches long and the other about 32, are fastened together to form a cross-shaped frame. A single piece of string is stretched around this frame to provide something to which the face of the kite can be attached. The face, consisting of a sheet of paper, catches the wind, enabling the kite to lift and stay aloft. The bow, formed by bending the cross piece, helps the kite to ride the wind by self-correcting its flight. The tail enables it to counteract the alternating wind currents. The kite is controlled while aloft by the line extending to the person flying it.

Example of Functional Analysis

To fly an Eddy bow kite skillfully, a person should know how it functions. Two thin strips of wood, one about 24 inches and the other about 32, are crossed and attached to each other, forming a cross-shaped frame. To make the frame, the longer piece is placed vertically and the shorter one is fastened horizontally to it. A notch cut about seven or eight inches from the top of the vertical piece forms a firm joint, holding the horizontal one in place. A continuous piece of string is stretched around the frame to provide something to which the face of the kite can be attached. This string is inserted in the slits in the four ends of the sticks to keep it from slipping off in the wind. The slits are reinforced by being wrapped with tape or short pieces of string. The face, diamond-shaped to fit the frame, is folded around the perimeter to form flaps. These are pasted over the string around the frame to hold the face of the kite in place. The face enables the kite to lift and stay aloft. The bow, in the horizontal crossmember, angles the sides of the face upward. The tail helps to keep the kite upright and steady by controlling its twisting, turning, and diving in response to sudden changes in wind currents. The kite is controlled by the line extending to the person flying it. By pulling and releasing the line, the kite's movements can be controlled.

Example of Causal Enumeration

An Eddy bow kite flies because of its aerodynamic design and construction. Its lightweight material causes it to be lifted by a breeze moving close to the ground at eight to twenty miles per hour. The design of the kite's face creates enough wind resistance to force it backward and upward. The bow lifts the kite's sides to cause the wind to flow at different speeds over its upper and lower surfaces. The bow also causes it to be self-correcting in flight. The tail produces enough air resistance to sudden changes in wind currents to stabilize the kite's movements. The line enables the person flying the kite to bring about changes in its elevation and direction of flight.

Example of Causal Analysis

An Eddy bow kite flies because of its aerodynamic design and construction. Its lightweight material causes it to be lifted by a breeze moving close to the ground at eight to twenty miles per hour. It would be too heavy for a slower breeze, and a faster one would cause the kite to spin and dive toward the ground. The design of the kite's face creates enough wind resistance to force it backward and upward. It is tipped forward by the line, held by the person flying it, causing the wind to push up under the tipped face to lift the kite, somewhat like a lever under a rock. The bow lifts the sides of the kite's face, causing the wind to flow at different speeds over its upper and lower surfaces. The air moves faster over the upper surface than over the lower. This faster-moving air creates a low-pressure area above the kite, and it rises into this low pressure in reaction to the high pressure beneath it. This is the same principle of aerodynamics that causes an airplane to fly. The tail produces enough air resistance to sudden changes in wind currents to stabilize the kite's movements. The Eddy bow kite has only one surface; it must have a tail to maintain balance. The tail's air resistance, not its weight, causes the stabilizing effect. The line enables the person flying the kite to change its elevation and direction.

Application 4–5

Read each paragraph carefully and state whether it is intended to inform or to persuade. Next, tell whether it is developed mainly by structural, functional, or causal enumeration or analysis.

1. Many of us who are in favor of proposals to govern the use and ownership of firearms should stop to consider some of the reasons why there is such controversy about them. Never has it been proven that gun control reduces crime in the United States, and there is hardly enough evidence for hope in that respect. Never have so many law-abiding people owned so many guns for recreational purposes. Never has the criminal not been able to obtain a gun anytime he wants one on the "black market." Therefore, let us take heed of these reasons and eliminate the threats to one of our greatest traditions, the right to keep and bear arms.

(Notice that the sentences in the series are similar in construction to indicate intentional enumeration and to gain emphasis.)

2. Blood performs transportation and sanitation functions that are essential for human life. It is a red sticky fluid circulating through the arteries,

capillaries, and veins. It is chiefly composed of plasma, red cells, and white cells. The plasma carries waste materials to the organs for excretion. The red cells, so small that a high-powered microscope is required to see them, give the blood its red color and act as oxygen carriers. The white cells, also small but not as numerous as the red cells, attack germs and act as scavengers. Blood is one of the body's chief protections against infection because it performs these important functions of transportation and sanitation.

3. The wind circulation around a hurricane consists of three parts: The Outer Portion, the Region of Maximum Winds, and the Eye. The Outer Portion is made up of wind that increases in velocity as it moves closer to the center. This region extends from the periphery of the storm where the wind may be no more than 20 to 30 mph to within 15 to 35 miles of the center. The Region of Maximum Winds surrounds the Eye of the storm. In this region, the wind velocity reaches between 120 and 150 mph. The Eye, the innermost portion of the hurricane, has a radius of 15 to 25 miles. The winds in the Eye are relatively constant.
4. What kind of man services your automobile? Undoubtedly, he is a dedicated auto mechanic, a trained, skilled tradesman. He must know front end alignment, the electrical system of your automobile, and how to use modern testing equipment to accurately diagnose your mechanical problems. Also, he must know the fuel system, including carburetion, fuel pump, gas filters, gas lines, and the gas tank. It is necessary that he knows how your automobile engine functions with all of its many working parts, with its complete driveline operation, with its automatic or manual transmission and with its differential operation, which enables you to go around corners. He must also know the brake system, whether your car has manual, power, or the new power disc brakes.
5. An audio system can be analyzed according to the function of its components. The basic audio system consists of a record, a tape recorder, a preamp, a power amplifier, and the speakers. The record player and tape recorder play their songs into the preamplifier. The preamplifier raises the output of a low-level source (the record player or tape player) so that it may be further amplified by a later stage. This later stage is known as a power amplifier. The power amplifier is designed to deliver a relatively large amount of energy in order to drive the speakers. The speakers are what give you your final output sound.
6. The Eye of the hurricane, its innermost portion, has a radius of 15 to 25 miles. The wind in the Eye is relatively constant. The Region of Maximum Winds surround the Eye, and its wind velocity reaches 120 to 150 mph. This region extends from the periphery, where the wind may be no more than 20 to 30 mph, to within 15 to 25 miles of the center, where the wind velocity ranges between 120 and 150 mph. The Outer Portion is made up of wind that increases in velocity as it moves closer to the center. These three main parts, the Outer Portion, the Region of Maximum Winds, and the Eye, make up the hurricane wind circulation.
7. A dance orchestra consists of fifteen to twenty instruments. The orchestra is separated into four main groups: brass, woodwinds, strings, and percussion. The brass group consists of trumpets, french horns, tubas, and

Enumeration and analysis compared

trombones. Clarinets, flutes, and oboes make up the woodwind section. The third group, strings, is comprised of violins and bass violins. The last group, percussion, is the rhythm section made up of the drums. These groups in combination blend the music to create the well-organized balance of a good dance orchestra.

Application 4–6 Read the paragraphs and tell whether each is enumeration or analysis and whether it is structural, functional, or causal. Explain why you think so for each.

1. The Welder is easy to use because of its extra features. Arc stabilization imposure of the high-frequency type is a feature of this machine, making it easy to strike and maintain correct arc temperatures and length without constant running to the machine to make small adjustments. The stabilizer makes AC rods more economical to use as well as better, sounder, and stronger welds. On the DC series, the range adjustment allows for low-amp welding, low weldment temperature, plus excellent penetration on straight polarity. On reverse polarity this machine can handle high-amperage jet and drag rods for high-rate filling jobs where deposit rate is important. The inert gases used with nonconsumable electrodes cause very sound welds on nonferrous metals.
2. What is high-frequency AC welding and why does it work? Starting with low-voltage high-amperage AC current would be a problem, because AC current flows both one way, then reverse, causing two instantaneous movements of no current flow per cycle of electricity. This presents a problem in welding because the arc of the electrode is hard to maintain. Now, if high-frequency, low-amperage AC current is imposed over existing AC high-amperage current, the many-times-multiplied number of cycles produces an even flow of current on the high-amperage use because the high frequency is overriding the low amperage and is flowing when the low amperage changes cycles, causing no shutoffs at cycle change. This greatly improves the stability of the arc to keep it from random wandering, causing the arc to go straight and deep.
3. The Welder is a quickly set up and easy-to-maintain machine, with extra features for universality. This welder uses only 14 square feet of floor space, making it a compact machine that can be located easily for on-the-job problems. This machine has double-wound coils, so if one burns out, the other takes over at half load, creating an economical yet smooth-running machine. It runs on a 40 to 500 amp DC range and has both CDRP and DCSP. Its AC range runs from 40 to 300 amps with a high-frequency overrider for stabilized arc. The cooling systems use oil-permeated water to cut water abrasion to almost nothing while providing lubrication to all flexible and moving parts. The water-permeated oil is pro-neoprene, to reduce stiffening and cracking in the hoses and seals.

Application 4–7 Read each of the following sentences carefully and determine whether each could be used more effectively to inform or to persuade. Now state whether each should be developed by enumeration or analysis to achieve that purpose best. Then select one for enumeration and another for analysis and write a paragraph for each.

1. One chair in our livingroom is constructed in an Early American design.
2. The atmosphere in an office is greatly influenced by its interior decorating.
3. Solving a problem scientifically involves only five simple steps.
4. It is easy to take a cigarette lighter apart.
5. Forest fires have three main causes.

Application 4–8

Select one of the items listed below and write a structural, functional, and causal analysis for it. Be sure to tell the purpose (to inform or to persuade) for each of the three and the kind of analysis you intended it to be.

1. The kinds of instrument in a band.
2. The buildings on the campus.
3. Athletic equipment.
4. Woodworking tools or cooking utensils.
5. Mouse trap.
6. Faucet.
7. Toaster.

Application 4–9

Study the accompanying drawing of a bench and pipe vise and write three separate paragraphs:

1. A structural analysis.
2. A functional analysis.
3. A causal analysis.

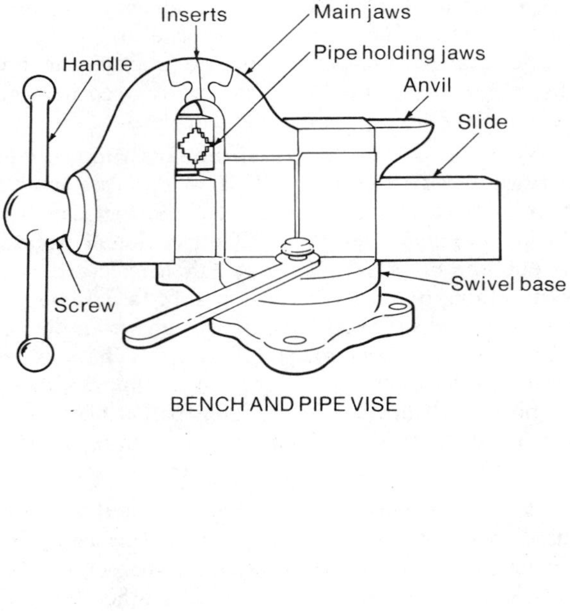

BENCH AND PIPE VISE

Application 4–10 Examine the accompanying drawing of a sink drain and write a paragraph or two of structural, of functional, and of causal enumeration. Tell which you intend each to be.

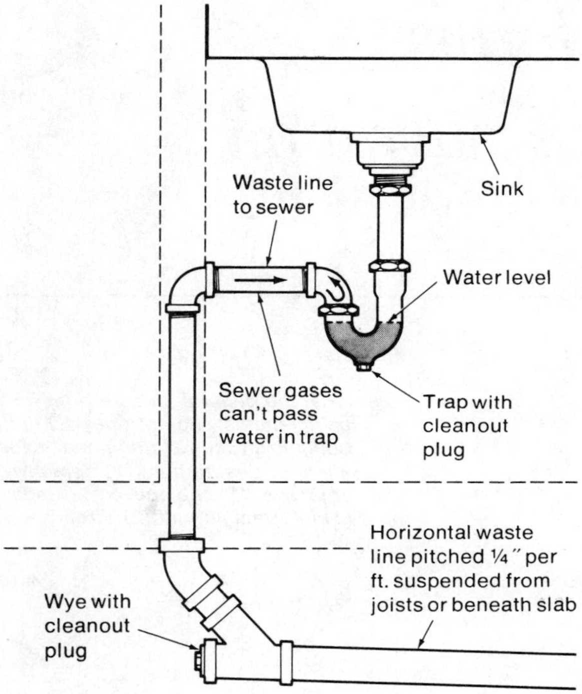

TYPICAL DRAIN LINE

Application 4–11 Write a paragraph or two of structural, of functional, and of causal analysis telling about a problem related to drain blockage of the sink illustrated in the drawing.

5

General forms: description, narration, causation

what the general forms are, how they differ . . . using enumeration or analysis to develop them . . . building structural, functional, causal relationships . . . kinds of description: (1) objective, (2) semi-objective, (3) subjective . . . kinds of narration: (1) informative and (2) directive

Description, narration, and causation are important forms of development in most kinds of communication, including speaking and writing about technical subjects. Sometimes the message is made up of details that mainly develop only one of these general forms. Often a combination of forms is used, but usually one of them is dominant.

In order to use description, narration, and causation in speaking or writing about your work, you should know what they are, how they differ, the kinds of details each requires, and how the ideas in them are assembled.

Because most communications consist mainly of description, narration, or causation, these may be called "general forms of development." For example, when explaining the architecture of the house in which you live, you may tell (1) what it looks like, (2) how it was assembled, or (3) why it was designed a certain way. The first will require description; the second, narration; and the third, causation.

All three of these general forms may be developed by enumeration or analysis. If you want to discuss the construction of a house, you may simply enumerate the features you detect about the house, the steps in the process of the roof assembly, or the reasons why the walls were designed that way. But if you want to develop these details by analysis, you must also explain the structural, functional, or causal relationships between them. Doing this will enable you to discuss the subject in depth in order to inform or persuade.

DESCRIPTION

Description tells what a person physically senses when experiencing something. When you describe, you express what you see, hear, smell, feel, or taste. You also may give some of your emotional reactions to what you experience.

The sense of sight plays an important role in description. Here are some of the visual details you can use in describing, for example, a house:

1. Form, shape, design (trilevel, bungalow, ranch).
2. Color.
3. Parts or sections (roof, windows, balcony, porch, kitchen).
4. Size (height, width, length).
5. Materials (brick, cement, wood).
6. Motion (moving parts: doors, windows).
7. Uses (sewing, storage, sleeping).

The following description of cigarette lighters in general is developed mainly by structural analysis. Can you see how the relationships between the parts are explained? Also, notice the details on motion and use. This description is a section, not a single paragraph. Two or more paragraphs of a certain form of development together make a supporting section, not a supporting paragraph.

> Most cigarette lighters are made with a few simple parts. The main parts fit inside a small case. Some cases are gold- or silver-plated. Others have decorative inscriptions on them. A few are even studded with diamonds.
>
> Inside the casing, a wad of cotton and an absorbent wick run up through the cotton to an outlet next to a piece of flint held firmly by a spring and screw. Next to the flint is a small abrasive wheel with coarse horizontal corrugations. When the user of the lighter presses down on a lever or button, the abrasive wheel rotates and rubs against the flint, causing sparks to fly outward. They ignite the wick, saturated with a highly inflammable fluid.
>
> Usually, each lighter case is equipped with a hinged hood. To extinguish the flame, this hood is brought down over the flame, cutting off the oxygen.

Viewpoints for Description

To write good description, you should decide in advance on the best viewpoint for the kind of response you want from your reader or listener. When you want to inform or to persuade, you have to know whether you should regard your subject objectively, semiobjectively, or subjectively. Review the discussion of these in Chapter 1.

When viewing your subject from a subjective viewpoint, your personal feelings will play a large role in determining what you see. As a result, your description may not be as accurate as one you may write from an objective or semiobjective viewpoint. Some subjective description is used when the writer or speaker wants to give a bit of interesting information or emotional appeal to persuade someone, as in advertisements and television commercials. It is often used in the public relations material that most large companies publish regularly.

Notice how personal the following subjective description is:

> The picture on our living room wall enchants me with curiosity, and lures me away from my chores to some small Pacific island. It depicts a long crescent of cinnamon-colored beach, lacing a tropical island cove. The bluish green water not ruffled by the surf is scattered in spots by the sun's rays into an assortment of blues, greens, and reds. Small rows of tall pines stand like sentinels as if waiting for the arrival of some expected dignitary. A sailboat, with its torn red sail and oar leaning against its bow, lies like a broken toy on the soft beach sand. The sun tinges the evening sky with dark orange as it sets.

Objective description is realistic description. It tells what something looks like as it really is. In a sense, it is a black-and-white photograph of the subject. The writer or speaker does everything possible to avoid expressing personal emotions. Special attention is paid to selection of the words to be sure they are not tinted or saturated with emotions. Here is somewhat the same picture referred to in the preceding illustration. This time, however, it is seen by the writer from an objective viewpoint.

> The picture on our living room wall is about 35 by 15 inches. It is surrounded by a walnut frame, which makes the painting stand out more. It depicts an island, seven palm trees, an ocean, a sailboat, and the sky. The island is all brown sand. The palm trees have green leaves and dark brown trunks. The larger trees are in front; they get smaller as if the viewer were looking into the distance. The ocean is a bluish green, and the sunrays make it lighter in places. Near the shore is a sailboat. Its red sail is torn and there is an oar lying across the bow. The sun is setting, and the sky is a dark orange.

Semiobjective description results when the description conveys a small amount of the writer's personal feelings about a subject. To keep the reader's interest it is often necessary to tint a description with personal emotions. In technical communication, you should try to avoid exaggeration in your semiobjective descriptions. Instead, select and use a few words that add a bit of feeling and color to keep the listener or reader interested in what you have to say. Here is the same picture described in the two preceding examples. This time, however, the paragraph is written from a semiobjective viewpoint.

> The picture on our living room wall is refreshing on a hot day. It depicts a pleasant small island with a few palm trees with long green fronds and deep brown trunks. They stand tall and straight, with the larger ones in the front row, adding an enjoyable perspective and balance to the scene. The bluish green of the ocean is in harmony with the color of the evening sky. A sailboat with its tiny red sail torn and an oar lying across it rests quietly on the shore. The setting sun is turning the sky into a dark orange.

Below is another good example of a semiobjective description. Notice how the italicized words color the description with the writer's personal feelings about the violin. Also notice how the amount of emotion is controlled to achieve the semiobjective, not the subjective, viewpoint.

> A modern violin displays precise craftsmanship. The *gently* curved back and belly are *painstakingly* carved from *seasoned* beechwood. The *frail* body is carefully glued to the long, narrow neck. The neck is partially made of ebony, shaped and sanded *with care* to sustain the rapid movements of the violinist's nimble fingers. The sturdy oaken pegs at the *scroll-like* head of the instrument are *attentively* polished and placed in their holes. Seemingly *fragile* catgut strings are *gingerly* wrapped in fine steel wire and wound around the pegs. Tautly stretched, the strings render *melodious* tones. The making of a violin today is a *meticulous* craft in which even Stradivarius would be proud to participate.

Did you also notice how this semiobjective description is achieved by structural enumeration, not by structural analysis?

Details and Sequence

The kind and amount of sense details you put into description depend upon how vividly you want your receiver to experience it. In addition to the visual features of the subject, you may add sounds, aromas, and touch details, such as smoothness, softness, or roughness. For some subjects, you may even want to add taste details. Even though some things can't be tasted, to express your feelings you can say things such as, "Operating the punch press was a sweet experience." This, of course, only tells how you feel about operating it.

Description must be expressed in an orderly manner. To maintain this orderliness, you should arrange details in a well-organized pattern in space. At times, you may need to make clear to your reader the distance and angle from which you are viewing the subject. Sometimes you may describe your subject in the order in which you normally view it in space. When you watch people, for example, you may observe them from their heads and faces downward to their shoes, or vice versa. Or your purpose might lend itself to describing first a front view, then a view from the rear. This orderly movement in space when describing may be used for any person or object.

Following is a supporting section of general description taken from a much longer report about the five-ton wrecker used in the United States Army.

> The wrecker is a six-wheel-drive tow-truck and mobil-crane combination with heavy winches at both front and rear. It is powered by a six-cylinder engine with a five-speed manual transmission that has both high and low range for light or heavy towing. The vehicle has a power takeoff for operating the winches and a selector to disengage the front wheel drive. It also is equipped with an air compressor for the air brakes.
>
> At the front of the vehicle, located in the bumper, is a winch assembly containing 200-ft of ½-in. stranded steel cable. On the rear of the truck bed is another winch, this one larger than the front winch, containing 300 ft of ¾-in. stranded steel cable placed evenly on the winch drum by means of a level-wind device.
>
> The main and most versatile unit on the truck is the boom-crane assembly, which is operated from a cupola located next to the main center pivot assembly. As the operator is seated in the cupola, he is facing the rear of the vehicle, and directly in front of him are four vertical hydraulic control handles. Reading from left to right the controls are: (1) Crane pivot: forward—left, back—right. (2) Boom elevation: forward—raise, back—lower. (3) Boom travel: forward—extend, back—retract. (4) Boom winch control: forward—lower, back—rewind. Care must be exercised when operating the hydraulic controls to gently activate the units and not pull controls hard, as this may damage the hydraulic system. The boom winch is equipped with 100 ft of ⅝-in. stranded steel cable and is capable of lifting 10,000 pounds. The boom assembly can pivot 170 degrees in either direction from center and can extend 20 ft beyond the sides of the truck. To prevent tipping, the vehicle is equipped with four outriggers located in the front and rear of the bed base.
>
> Miscellaneous equipment carried on the vehicle includes several heavy and medium chains of various lengths, an assortment of wrecking bars and tools, an acetylene torch and its accompanying accessories. Also included are large heavy tow ropes and three or four sheave snatch blocks for mechanical leverage when winching.

Application 5–1 Write a supporting section of at least three paragraphs describing the cabinet that is illustrated here. Develop each paragraph by structural enumeration. You may add your own choice of finishing, trim, and other embellishments in your description.

PORTABLE BASE CABINET

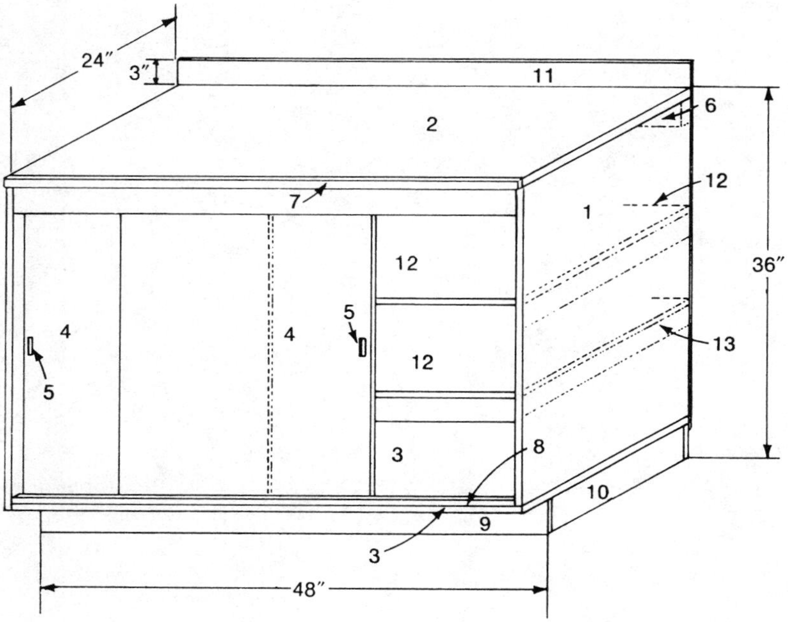

Number of parts	Name of parts
1	Ends (2) 24″ x 31¼″
2	Top (1) 24″ x 48″
3	Bottom (1) 24″ x 48″
4	Door (2) 24″ x 30″
11	Back (1) 36″ x 48″
12	Shelves (2) 21¾″ x 46¾″
13	Shelf supports (4) 2¼″ x 21½″
5	Pulls (2) ¾″ x 3″
6	Support top (2) 1″ x 4″ x 46¾″
7	Front trim (1) 1½″ x 48″
8	Door guide (1) ¾″ x 46¾″
9	Front toe (2) 1″ x 4″ x 48″
10	Side toe base (2) 1″ x 4″ x 19½″

Application 5–2 Write a supporting section of at least three paragraphs describing the rolling picnic table shown in the accompanying drawing. Develop each paragraph by structural, functional, or causal analysis.

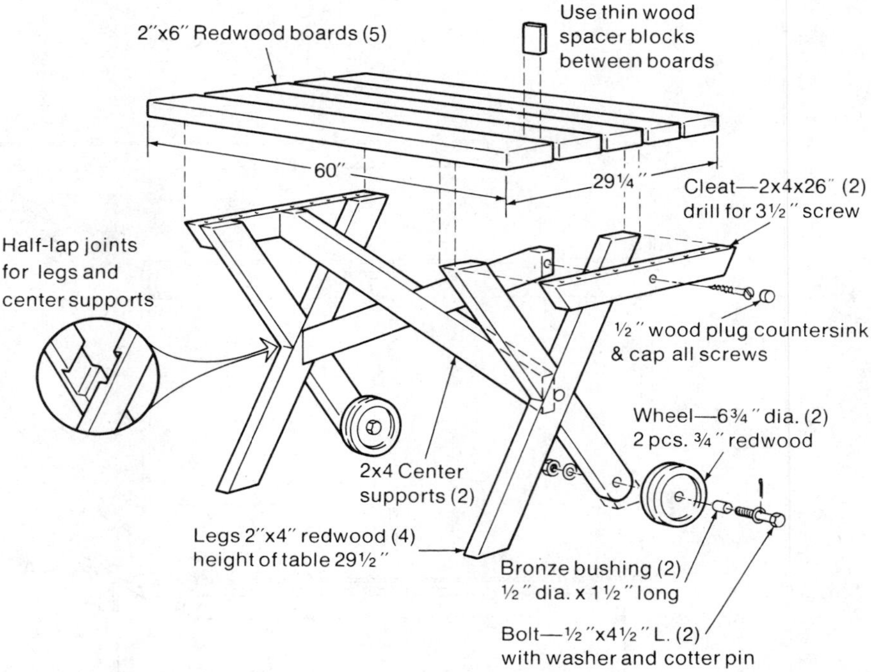

Application 5–3 Carefully read the "Description of True Hickory" below, and write as many as you can of the corresponding details you find in the article under the following categories:

1. Form, shape, design.
2. Color.
3. Parts or sections.
4. Size.
5. Materials or ingredients.
6. Motion or changes.
7. Uses.

DESCRIPTION OF TRUE HICKORY [1]

A tree in the true hickory group is a medium-sized, deciduous hardwood that grows in the humid climate of the Eastern United States. It is generally slow-growing, but does respond favorably to increased soil fertility and has a reputation for long life. Some exceptional trees have

---

[1] Douglas R. Phillips, "Hickory," U.S. Department of Agriculture, Forest Service (Washington, D.C.: U.S. Government Printing Office, 1973), p. 1.

Description

been known to live 350 years. The true hickories vary in size by species and location, and some trees attain heights of 120 to 140 feet and diameters of 36 to 48 inches. The trees are shade-tolerant, highly competitive, and reproduce readily from both seed and sprout.

The leaves of true hickory are compound, 8 to 14 inches long, and have 5 to 9 ovate-to-obovate leaflets. The leaflets are oppositely branched and are characterized by an aromatic odor when crushed. The fruit is a round or egg-shaped nut 1 to 1¾ inches in diameter. It is covered by a hard husk, ⅛- to ¼-inch thick, which splits lengthwise along the seams of the husk. The meaty part of the nut is edible, although it is sometimes bitter (fig. 2).

The size of the seed, or nut, varies considerably between species. Shellbark has the largest seed (25 to 35 seeds per pound), pignut the smallest (approximately 200 seeds per pound). Mockernut averages 90 seeds per pound and shagbark 100 seeds per pound.

The bark of young hickory trees is characteristically smooth, gray, and very hard. As the tree grows older, the bark either breaks up into plates or forms ridges which are separated by narrow fissures. Shagbark develops plates that are wide and curled at the ends so as to give a shaggy appearance (fig. 3). Shellbark is similar to shagbark except that the plates are thinner and more firm. Mockernut and pignut have furrowed bark with interlacing ridges. Mockernut has firm, almost horny ridges whereas pignut has somewhat scaly ridges.

Seed production varies between species, but normally the true hickories have a good seed crop once every 2 to 3 years with light seed crops in intervening years. Approximately 50 to 75 percent of all seed produced are viable. The seed-producing years are from age 25 to 200, with optimum production occurring from age 40 to 125. The rate of production is important for squirrels, chipmunks, raccoons, and other small animals since hickory nuts provide a large part of their food supply. These animals are credited with the major portion of seed dispersion for the species.

The wood of the true hickories is known for its strength. Some woods are stronger than hickory and others are harder, but no other commercial species of wood is equal to it in combined strength, toughness, hardness, and stiffness. As a basis of comparison, hickory is approximately 30 percent stronger than white oak and 100 percent more shock-resistant.

In cross section the growth rings of hickory are very distinct, as are the heartwood and the sapwood. The heartwood is pale brown to reddish brown, the sapwood is white and approximately 2 to 4 inches wide in a tree 12 inches in diameter. The wood is straight-grained but has a coarse, nonuniform texture.

Hickory is used to a limited extent in a variety of products which include flooring, veneer and plywood, railroad crossties, fuelwood, and charcoal. Hardwood flooring seems to be the most dominant of these. In 1965, 16 percent of all hickory used went into flooring, an increase of 8 percent in a period of 17 years. Hickory veneer has been produced on a small scale for use in paneling, specialty items, and a small amount of plywood. The major drawback in producing a large quantity of veneer

General forms: description, narration, causation

appears to be the lack of veneer-quality logs. The tendency of hickory to check during drying and the difficulty of treating it with a preservative have prevented greater use of hickory in the production of crossties. Hickory is very popular as a fuelwood and as a charcoal-producing wood. The fuelwood burns evenly and produces long-lasting, steady heat; the charcoal gives food a hickory-smoked flavor.

Application 5–4

Select a topic from the list below and write a section of description, consisting of at least two paragraphs. Be sure to identify your predetermined purpose—to inform or to persuade—before you start, and tell which it is in your heading. Also, tell whether you developed each paragraph by enumeration or analysis and whether it has structural, functional, or causal relationships.

photograph	machinery
desk	cooking utensil
building	furniture
clothing	laundry equipment
windmill	kitchen appliance
aquarium	

NARRATION

On the job, you often need to tell a story or listen to one to do your work. You have to tell someone how you did something or listen to directions as to how something should be done.

Narration tells a sequence of related happenings explaining how something originated, occurred, or is done. In narration you may use mainly functional enumeration or analysis. Focus the attention of the reader mainly on the happenings. Some description will be needed, but the main attention should be on the series of incidents or steps. When you explain the happenings, be sure to move through time in an orderly manner.

Types of Narrative

In communication, either of two kinds of narration is used, climactic or straight-line. *Climactic narration* is seldom used in technical communication. It is the kind used in short stories and novels, in which the suspense is increased in intensity to a snapping point or climax.

Straight-line narration, the kind usually used in technical communication, simply tells what happened or how something is done, concentrating on information rather than emotional effect; there is little or no variation in intensity. When you tell a person how to get a driver's license or how to find a certain store, you are using straight-line narration. The history of a business and a résumé or personal data sheet are straight-line narratives. The data sheet is a form of autobiography. Both biographies and autobiographies are straight-line narratives.

Process and procedural narratives are the most common types used in technical communication. These tell step-by-step how something is done or should be done, such as how to bake a pie, paper a wall, or drive to Cobo Hall in Detroit, Michigan. . . . They also often tell how something happened—a traffic accident, the eruption of a volcano, or the emerging of a butterfly from a caterpillar.

Examples of Narrative Sequence

The following is a brief supporting paragraph taken from a longer piece of historical narration. It tells the sequence of steps involved in the discovery and development of electric batteries.

> The first true electric battery, called the voltaic pile, was discovered by Alessandro Volta in 1800. Until then, the Leyden jar with its static generator had been the only worthwhile source for electricity. With Volta's discovery, the science of storing electrical current took a great step forward. In the early days of electrical science, primary batteries were the main source for direct current. From 1800 to the early part of the 20th century, wet batteries were used. The dry cell has replaced the wet type, which used sal ammoniac as an electrolyte. The copper oxide caustic soda cell is still in use for heavy duty service.

The narration that follows merely tells how something occurs or happens. Although there is description in it, the sequence of steps is the main form that holds it together.

> A reservoir of water through the miraculous nature of water seeks its own level. A well newly dug is an empty chamber, a deep pit in which the walls change from top soil to clay as it deepens. Beads of water soon form on the sides, like sparkling diamonds studding the walls. First one drip of water, then another, until hundreds stand out all around. They begin to trickle down to the spot where a droplet first started out. With the addition of other droplets, together they form a small stream. Soon there are many little streams which slowly merge, decreasing in number and increasing in size. These streams continue to grow in size and become the main channels of water supply which forms a deep pool of clear, cool water at the bottom of the well.

You may never want to make fabric from cotton, but you might want to be informed about how it is done. The following is specific enough to do just that.

> Raw cotton goes through a long series of processes before it is finally woven into fabric. The cotton is first cleaned and formed into large rolls. These rolls are fed into machines that straighten and smooth the tangled fibers. This process is called carding. Now the cotton has been formed into a ropelike strand. These strands will be combed to straighten out the fibers. To make this strand into yarn strong enough for weaving, it must be pulled and twisted so that the fibers form a tight thin thread. The pulling or drawing out is done on a drawing frame. This is repeated

several times, drawing out the fibers into finer and tighter strands. The cotton is finally spun into yarn on spools called bobbins. These bobbins of yarn are then taken to a weaving room where the cloth is made.

Below is a good section of narration. It tells how something occurs in a way that would interest a person with only a bit of knowledge about insect-eating plants. To do this, the writer allowed some words that indicate emotion to color and add some warmth to the information. Some of these are printed in italics.

> Meat-eating plants are *not easy to find*, but the *clever* traps they have invented to get the meat they need enables them to survive almost everywhere. The *tricks* these meat-eaters use to *ensnare* their prey are *amazing*. Although each has its own invention for trapping insects, they all have pretty much the same kinds of enzymes and acids in their stomachs with which to do the job of digesting them.
>
> The most common type of trap is used by the pitcher plants. Pitcher plants, also called trumpet flowers, like lilies, have pitfall type traps. A trap has a one-way passage that leads right to the tip of a pit. Once the insect is *enticed* that far, sharp, spikelike bristles, turned in at the end, keep it from escaping. Tempted by *delicate* aromas, the victim is *lured* on until it suddenly plunges into the pit. It then is quickly drowned in a pool of acid and slowly digested. Some pitcher plants have another feature, a lid that shuts and holds the prey.
>
> Because ants are so skilled at avoiding traps, but *so delicious*, some plants have invented special traps to capture them. Because ants can walk upside down so well, some Brazilian plants have had to develop the lobster-pot trap. The *prettiest* of these looks much like a *delicate wine flask*. A *spiral staircase* extends down the plant neck, growing smaller and narrower as it winds down toward the trap, baited with *appetizers* an ant *can't resist*. The ant follows the *aroma* until the passageway is so narrow it can't turn around. As it looks around to try to find a way out, it steps on a treadle, and the firm bars behind it snap to lock it in. It is then consumed slowly.
>
> Butterwort plants have developed their own *ingenious* way of trapping their meat. The trap is a kind of "flypaper." Butterworts lie flat on the ground with a glitter on their leaves that looks like honey. When the feet of a victim touch a leaf, the plant *lets go* with a sticky glue to hold the prey. Next, a shot of acid with digestive enzymes that overpowers the insect is spread over it. Finally, the edges of the leaf roll like a curtain over the victim to smother it.

The following is a good example of a narration telling the reader how he or she should do something. It provides the step-by-step sequence. It also explains the reasons for certain steps and carefully points out the precautions that must be taken.

> Repairing a dent in an automobile with body filler must be done quickly, yet carefully. Your first step is to sand away the paint and dirt from

Narration

the surface of the area to be filled, enabling the body filler to stick well. When the surface is smooth and clean, carefully apply the body filler in the dent. Quickly smooth it out in order to save time and energy when finishing. (Note: many kinds of filler dry and harden quickly, therefore, smooth out the filler soon after application.) The filler generally hardens within four hours (depending upon weather conditions). When it does, it's time to sand it down for smoothness, making it flush with the surface of the car body. When this is completed, a primary paint is applied. After a coat of primer is given, the finished color is placed over the filler making it "good as new."

Application 5–5

Select one of the following topics, predetermine your purpose, and write a section of narration consisting of at least two paragraphs. Tell in your heading what your purpose is, to inform or to persuade.

How a traffic accident happened.
How a fire was extinguished.
Cooking something.
Doing a certain job.
Solving a problem.
Finding a certain place.

Application 5–6

Tell which kind of narration each of the following paragraphs is, and explain why it is mainly narration rather than description.

1. The search for a polio vaccine was long and hard. The first significant advance in research occurred in 1868 with Karl Landstiener and Erwin Prapper. Experiments were confined to monkeys for several years. In 1910 it was discovered that serum could be made from monkeys that had recovered from polio. But when used on humans it encouraged meningitis. As soon as one vaccine was invented, it was discredited and replaced by another. There were times when experiments would lead to several deaths, especially of children. It took until 1949 before it was discovered that not only humans' but monkeys' tissue formed a good culture to be used in vaccine production. In the same year, it was also discovered that there were three different types of polio virus. Finally, in 1954, the Salk vaccine was discovered.

2. True moss is formed by one of three different processes. Some types contain both female plants and male plants. Under certain climatic conditions the sex organ of the male plant will rupture and scatter its sperm. Since moss generally grows in a moist atmosphere, these sperm cells are propelled by their whiplike tails through water to the female plant. In other types of moss, the male and female sex organs are contained in the same plant, in which case the sperm cells travel through a canal to impregnate the ovum. The third process, which may also occur in either of the preceding types of plants at certain stages of their development, is called spontaneous regeneration. By this process the plant reproduces itself from small sections of its branches that break off and fall into fertile soil or water. From these separated sections, new plants grow and spread.

Application 5–7 Write a specific directional narration telling how to tie one of the knots in the accompanying illustration—the overhand knot, half hitch, or two half hitches.

a
b
b
a
Overhand knot
Half hitch
Two half hitches

Application 5–8 Study the accompanying drawing and write a section of narration telling someone how to sharpen a chisel. Be sure to do any needed research to enable you to do a good job. Be sure also to give explanations as to why something is done. Also, don't neglect to give any necessary precautions.

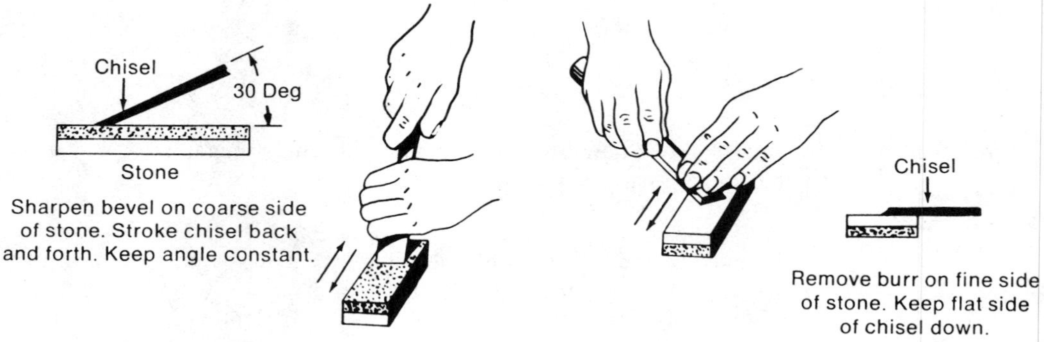
Chisel
30 Deg
Stone
Sharpen bevel on coarse side of stone. Stroke chisel back and forth. Keep angle constant.
Chisel
Remove burr on fine side of stone. Keep flat side of chisel down.

CAUSATION

Everything has a reason or a cause for its happening or existing. Development of the type known as *causation* may identify the causes, or it may tell about the search for a cause and report your findings and conclusions. Although it may be developed by structural or functional relationships, it is mainly developed by causal enumeration or analysis.

Here is a section of causation, consisting of several paragraphs, explaining why steel must have different degrees of hardness and how various kinds are obtained.

> The most important characteristic of steel is its hardness, which gives it its strength. Since almost every product manufactured from it needs a steel with a different degree of hardness, this quality must be regulated. This is done by controlling the amount of carbon content along with its smelting and quenching temperatures.

The first of these and the most important is quantity of carbon. By varying the amount of coke mixed into the ore during the smelting process, the carbon content is regulated, and, consequently, the steel hardness is made to comply with specifications.

Regulating smelting temperatures for steel is another reason for its degree of hardness. This temperature must be high enough to cause the carbon to dissolve, enabling it to disperse evenly throughout the molten metal. If this does not happen, soft pockets of steel will result. Controlling the steel temperature also controls the size of the steel grains. Very high temperatures will cause large-grained steel. This type is not as hard nor as strong as steel of smaller grains.

Varying the quenching temperature of steel also affects its hardness. This is the temperature at which the steel is suddenly cooled after it has been poured into the molds. This quenching temperature is predetermined by the degree of hardness needed for a certain application.

The following single supporting paragraph is developed by causal enumeration, listing a series of causes explaining why Colorado is a good place for business and fun:

Favorable weather conditions make Colorado suitable for recreation and for business as well. Power failure caused by severe weather is almost unknown to occur there. The majority of the highways in Colorado are open to traffic every day of the year. Stapleton International Airport is closed only an average of 13 hours annually; it is open 99.86 percent of the year. Skiing resorts are open in the high country from December to May. At most country clubs surrounding the Colorado cities, golf is played 300 days a year. Summers are comfortable even without air conditioning because of the low relative humidity.

The supporting paragraph below is causation. It explains the reasons why "ghosts" appear on a television set. It is different from the preceding paragraph because it is developed by causal analysis rather than by causal enumeration. Closely examine it to see how the writer doesn't just list the reasons for the ghosts but also carefully explains the causal interrelationships between them.

"Ghosts" or multiple images on a television screen are caused by reflected signals. Television signals travel in a direct path from the transmitting antenna to the receiving antenna on the television set. These signals are then converted by the television apparatus into images on its screen. Sometimes, however, the receiving antenna also receives signals that have been transmitted by the television station and then are reflected from something that gets in the way, such as a water tower, mountain or hill, airplane, etc. Since the reflected signals travel a greater distance than the direct signals, they are weaker and arrive later at the receiving antenna. Because these reflected signals arrive later, their timing is different from that of the direct signals. This difference causes the television set to produce a separate image or images. These separate images appear to one side of those produced on the screen by direct signals. They are weaker and give a "ghostlike" appearance; consequently, they are called "ghosts."

Causation is very important because it plays a dominant role in any communication intended to persuade. It is used whenever the intention is to cause the other person to discard a belief and to accept that of the speaker or writer. You should remember, however, that causation also is often used to inform. You will use it when you want to inform another person of the causes of something, perhaps an accident.

Since causation plays a vital role in interpretive technical reports, you will study its application again. In Chapter 15 you will be shown how it is used to write investigation, evaluation, proposal, and feasibility reports.

Application 5–9

Explain why the following paragraph is mainly causation rather than description or narration. Next, list the causes and reasons supporting its conclusion and tell whether the paragraph is developed by enumeration or analysis.

> A closed fabric drapery at a window is, by and large, an averager of the temperatures on its two sides. In cold weather, when the window glass is chilled, a closed drapery will, therefore, reduce the radiative heat loss of people sitting near a window, enough to improve their thermal comfort appreciably. It is true, however, that most draperies are not fitted tightly to the window, and room air can and does move quite freely by convection into and out of the space behind the drapery. For this reason, closing a drapery at night provides only a minor reduction of the winter heat losses of a house—considerably less than the reduction obtained twenty-four hours a day by using storm windows or a double glazing. However, even with these in use, there will be more comfort for people near windows if draperies are closed in cold weather. It should be noted that this improvement in comfort applies also to families living in apartments which might not be equipped with storm windows.

Application 5–10

Select an idea from the following list, decide whether you want to inform or to persuade, and write a section of causation consisting of at least two paragraphs. Be sure to indicate your intended purpose in your heading and tell whether each paragraph is developed by enumeration or analysis.

The causes for an accident or some other misfortune.
The reasons for cloud formations or other weather conditions.
The causes for volcanic activity.
What causes a match to light.
Why people buy a product.

Application 5–11

Following is a series of supporting paragraphs developed by description, narration, or causation. Identify each paragraph and tell whether it is developed by enumeration or analysis.

Causation

1. Experts consider the soybean to be the most useful vegetable in the world. More soybean oil is produced in the United States than any kind of vegetable oil. It oils machinery, goes into candles and soap, and is used in making oil cloth, linoleum, paints, and varnishes. Soybean plastic has been spun into fiber that makes an excellent artificial wool. The bean helps to make artificial rubber, explosives, insecticides, and adhesives that are valuable in making plywood. Soybean plants are excellent food for livestock and make fine silage. One pound of soybean flour has twice as much calcium as a pound of milk and was used in K-rations during World War II.
2. My desk lamp is well designed. It has a square base, which is tan colored. The base holds two pens and has a red button for "on" and a black button for "off." Protruding straight up from the base is a gold-colored curved bar from which the light hangs. A flourescent bulb encased in a metal shield reflects the light down on the desk. This metal shield is on the end of a swivel attached to the gold-colored bar, enabling the light casing to move up and down or sideways. The whole lamp, which stands about 15 inches high, throws a soft light over the desk, enabling me to work for hours without suffering eye fatigue.
3. Before beginning work on the ten-key adding-listing machine, you have to review the operative parts of the machine and the operational keys. Place your machine at a slight angle to your right; materials directly in front of you. This makes it easier to see your work and operate the machine. Before you begin a problem, press the total key to make sure the machine is clear. Note that the 4-5-6 keys serve as the "home" position for the fingers. Since the ten-key adding-listing machine is correctly operated by touch, you are urged to concentrate on developing the touch skill. Tap the keys in the same order that you read the digits from left to right. Strike the keys one at a time. Tap the 4 with the index finger of the right hand, the 5 with the middle finger, the 6 with the ring finger. Then tap the plus (or minus) bar to record the number 456 on the tape. After entering all items in a series, tap the total key with the little finger to total and clear your registry for the next calculation.
4. There are several main causes for abnormal tread wear on car tires. The wheels could be out of balance, which will cause irregular wear. Improper toe-in or toe-out of wheels also will cause irregular tread wear. Or the tire wear could be caused by improper camber (wheels tilted excessively inward or outward), causing more wear on one side of tires. Faulty or "grabby brakes" can cause the same conditions as out-of-balance wheels. The last reason for irregular tire wear could be faulty or worn shock absorbers.
5. The starting procedure of your outboard motor is easy after a little practice. Always make sure the shift lever is in the neutral position so you can't accidently start the engine in gear. Turn the speed control to the position marked start. Pull out the choke lever and then rapidly pull the starter cord until the engine starts. Once the engine is running, push the choke lever in and wait about two minutes for engine to warm up. After the warm-up you are now ready for the excitement of boating.

General forms: description, narration, causation

6. Assembling a dress using the new easy-sew pattern can be simplified with the new sizing material. The detailed instructions indicate how to lay the pattern using the minimum amount of material and alterations. After cutting each piece carefully, begin by sewing the darts on front and back sections. Next, apply back zipper and sew back seams. Match side seams and sew to complete the body of the dress. The neckline facing should be sewn to the outside of the neck and turned under, and then the sleeves should be set in to complete the garment. Next, try the dress on and pin the hem to desired length for the finished look. Anyone can sew using the new easy-sew patterns.

Application 5–12

Study the details of the accident shown on the accompanying drawing and write an accident report. First write a paragraph or two giving the basic details, such as the names of the drivers, the occupants, the directions the cars were traveling, the weather conditions, and the names of the streets. Then write a paragraph of description about the extent of the damages. Next, write two or more paragraphs explaining how the accident occurred. Finally, write a paragraph or two of causation, telling what conclusions you and the witnesses arrived at about the causes of the mishap.

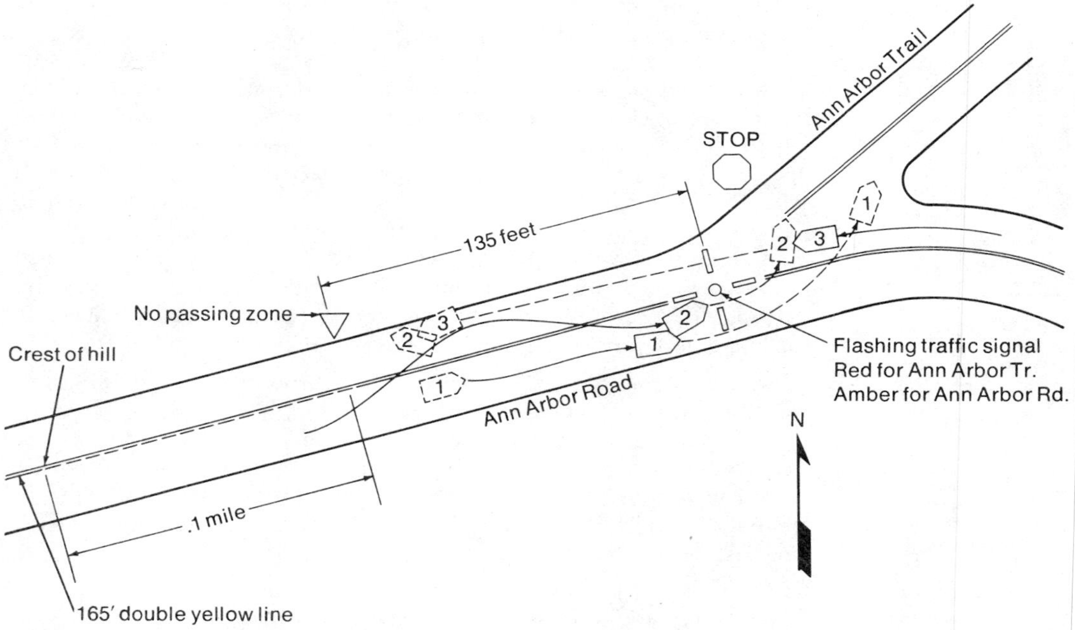

Application 5–13 The accompanying drawing illustrates how to correct a loose hinge. Very briefly identify the cause of the problem and write a paragraph or two of narration telling how it might be corrected with an empty toothpaste tube.

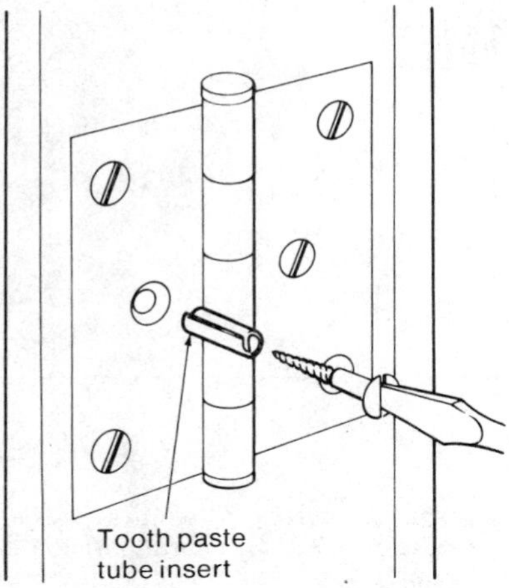

Supporting forms: classification and definition

6

what supporting forms are . . . kinds of relationships: structural, functional, causal . . . how to use them to inform and to persuade

In Chapter 5 you learned how enumeration and analysis are used to develop description, narration, and causation. The supporting forms shown below are used to expand, amplify, support, and reinforce these three general forms.

classification	comparison
definition	contrast
illustration	analogy

In this chapter we will be concerned mainly with classification and definition. The others will be discussed in Chapter 7.

You use these supporting forms whenever you say something. They are helpful when you want to be sure your listener understands complex ideas, especially technical ones. Since writers are not in direct contact with their readers, they can't use limb gestures, facial expressions, and voice manipulations to express exactly what they have to say, so they have to rely more on the supporting forms.

Here are some uses for these supporting forms in technical communication:

1. To adapt the technical ideas to your readers' educational and occupational experience.
2. To make objects and happenings more vivid and understandable.
3. To offer supporting illustrations as evidence in proof.
4. To sort and separate technical ideas for orderly development.
5. To explain clearly and vividly what technical terms and principles mean.

Although you may not be conscious of it, you use classification and definition every day to help you say something clearly in spoken form. You use them with little planning, almost automatically. In written communication, however, you will have to learn to use them deliberately and carefully, especially when writing about technical ideas.

CLASSIFICATION

Classification is sorting or placing something into a group, class, or category with other things with which it has one or more characteristics in common. In other words, classification is just sorting people, objects, places, ideas into different groups or classes, depending upon a quality or characteristic that all of the members of that group have. One common way of classifying people is by blood relationship. For example, to say that "Hank is my brother" classifies Hank as belonging to a certain family whose members have the same blood relationship.

Almost everything we do requires some kind of classification. We spend a large part of each day classifying. We classify our clothing, depending upon the kind of material, the purpose it serves, how much it cost, etc. In our dressers we place our socks in one pile, our handkerchiefs in another. In the kitchen we sort our food according to its perishableness, its packaging, or size. We even classify the people we meet by their physical appearance, their personalities, or the work they do: plumbers, electricians, doctors, lawyers, etc.

Whenever you give something a name or identify it by using its name, you are classifying. When you call something a "microphone," you are classifying it as an instrument that converts acoustical waves into an electric current, often fed into an amplifier, recorder, or transmitter. In this classification, the microphone is classified generally by what it is, "instrument," and more specifically by what it does, "converts acoustical waves into an electric current."

Every mind involved in a technical communication, whether it be that of the sender or receiver, requires that ideas be presented in an orderly manner. Classification is one of the basic ways by which this orderliness is maintained. Without it, a speaker's or writer's ideas would be a confusing hodgepodge.

How to Develop a Classification

The following procedure can help you develop a classification:

1. State the exact name of what you are classifying. If you think it necessary, define the thing you are classifying.
2. Clearly identify the basis for your classification, your standards, if necessary. Sometimes the names of the classes imply the standards, as when cars are classified by size: midgets, compacts, intermediates, etc.
3. Give the names of the classes into which you are sorting your subjects.
4. Briefly discuss the items placed in each class.

Figure 6.1 classifies vehicles by sorting them out into subclasses.

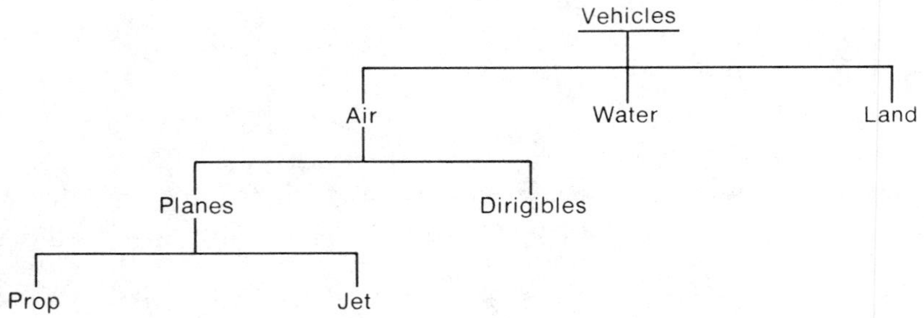

Figure 6.1
Classification of vehicles.

The feature or criterion upon which the classification is based is usually called the "standard." If, for example, you are going to classify automobiles according to cost, the standard would be the price or the price range. Consequently, you would place the Pinto in a class with others of the same price range; the Vega in another; the Impala, in still another, and so on, depending upon how much each costs. You could classify these same cars on the basis of other standards, such as body design, size of engine, or wheel dimensions.

Classification Standards

Any item or idea may be classified in a variety of ways because of the assortment of standards that may be used. Three general types of standards you already know something about are related to the kinds of relationships in enumeration and analysis. These are structural, functional, and causal standards.

1. *Structural standards* classify something by its components, design, and other physical features.
2. *Functional standards* classify something by how it works, what it does, or how it does it.
3. *Causal standards* classify something by the kinds of causes or reasons that explain why it exists or happens.

When you classify a person as a boy, for example, you are classifying him by structural standards, and all persons in that class have certain structural features in common. To say that boy is a football player, however, classifies him functionally, by something he does. When you classify the same boy as a pneumonia patient, you are classifying him by the cause for a physical disorder.

To illustrate classification based upon structural standards, here is a supporting paragraph classifying automobiles. The standard used here is dimension, which is subdivided into three classes. The term, standard, and classes appear in italics.

> *Automobiles* [term] fall into one of three classes depending upon their *dimensions* [standard]: *small, intermediate,* and *large* [classes]. The small car ranges in length from 112 to 157 inches and weighs from 1,300 to 2,500 pounds. Usually it is powered by a four-cylinder engine. The intermediate car ranges in length from 203 to 210 inches and weighs anywhere between 3,500 and 4,000 pounds. It is often equipped with a six-cylinder engine. The large car varies in length from 225 to 230 inches. Its weight ranges from 4,500 to 5,500 pounds. It requires a larger power source, an engine ranging from 300 to 420 horsepower.

The next illustration is a section of classification based on functional standards. The Japanese fishing industry is sorted out, based on where and how the fishing is done. Where the fishing is done as a part of the standard is clearly indicated by the class names.

FISH HARVESTING IN THE FUTURE [1]

> The seas surrounding Japan have always been rich in all forms of marine life, and the Japanese, since antiquity, have taken a substantial proportion of their food supply from these fertile fields. Thus, Japan today is one of the major fishing nations in the world.
>
> The total Japanese fishing catch in 1968 was 8,670,000 metric tons, representing 13.5 percent of the total world fishing haul. The industry can be divided into three broad categories: coastal fishing, offshore fishing, and pelagic or deep-sea fishing.

---

[1] Basil M. Mailey, "Fish Harvesting in the Future," *The Fisheries of North America*, U.S. Department of the Interior, Bureau of Commercial Fisheries (Washington, D.C.: U.S. Government Printing Office, 1966), p. 38.

Coastal fishing is conducted either by boats of less than 10 tons, by set nets, or through artificial breeding techniques in shallow waters, popularly known as "fish-farming." Although a large number of persons are engaged in coastal fishing (557,000 persons in 1968, or 93.7 percent of the entire industry), its production amounted to only 23.6 percent of total value. As a result, coastal fishing productivity is about one-fourth that of the larger fishing enterprises, and more recently, fishing grounds for the coastal fishing interests continue to shrink because of water pollution by waste water from factories along the nation's coastal waters.

Pearls, oysters, etc., have so far been cultured in shallow sea waters. In addition, yellow-tail, prawn, and octopus are now being raised by the shallow sea culture method and those thus grown are seen on the market with increasing frequency. The rise in the total value of fishery production, which increased by about 55.4 percent between 1963 and 1968, is attributed mainly to the rapid development of this shallow sea culture program, which has been promoted by the government.

Offshore fishing is conducted mainly by medium-sized enterprises with boats in the 10- to 100-ton class. The offshore fish catch amounts to about 36 percent of the nation's entire catch, but in terms of value it amounts to only 24.5 percent of total value.

Pelagic or deep-sea fishing is engaged in by large fishing vessels which operate in waters far from Japan. Their catch amounts to about 32.6 percent of the total catch. Trawling centering on the water off the African continent is conducted by ships of the 1,000-ton class. Tuna fishing in equatorial and subequatorial waters around the world is conducted by ships of the 200- to 500-ton class.

Salmon and crab fishing in north Pacific waters and whaling in the Antarctic Ocean and the north Pacific are carried out chiefly by large fishing fleets with mother ships which act as large floating fish-processing and canning plants.

The following is a supporting paragraph developed by causal classification. "Desert bighorn competitors" are classified here by the extent to which they use up the food and water the sheep need to survive.

Desert bighorn competitors that cause a decrease in the number of surviving sheep fall into three groups, depending upon the amount of food and water each uses up. These are: natural competitors, introduced competitors, and man. Rabbits, rodents, and deer are examples of natural competitors, eating the same foods the bighorns eat. Generally, deer competition is not excessive, but where it exists it most often occurs in the vicinity of water. Competition from introduced animals is much more important and serious. Included in this class are cattle, sheep, wild goats, burros, and horses. The livestock often concentrate near water sources with resultant overuse of forage in the area. These introduced competitors are often related to human encroachment. Farms, roads, fences, dwellings, and recreational areas usurp bighorn habitat.

When the standards are clearly implied by the names of the classes, it would be repetitious to state them. When you classify arrows as cedar, fiberglas, or aluminum, for example, you may not want to state the standard because it is clearly implied by the names of the classes. It is based on the construction material, a structural standard.

It is worth pointing out again that a supporting unit of classification may be more than a single paragraph. In fact, the whole report or article may be nothing more than classification. A supporting section of the type that follows is often called a classified summary, because the series of paragraphs summarize a series of classes.

> Size, shape, and quality are used by commercial breeders as a guide in classifying cattle into any of three groups. The three major classifications are: (1) beef cattle, for their meat, (2) dairy cattle, for their milk, and (3) dual-purpose cattle, for their meat and their milk.
>
> Beef cattle are larger and heavier than dairy and dual-purpose cattle, and they usually mature faster. These cattle are squarely built, with thick, short bodies, broad, deep chests, and short, muscular necks. They have short legs with plump thighs, and their backs and bellies are almost straight.
>
> Dairy cattle, raised mainly for milk, are usually smaller and leaner than beef cattle. Most dairy cattle have long, narrow heads, long necks, straight backs, and flat, lean thighs. They are characterized by large utters that are long and wide.
>
> Dual-purpose cattle are usually raised on small, nonspecialized farms or in areas lacking land and climate for raising dairy and beef cattle. These usually produce less milk than the good dairy cattle and their meat is inferior to that of the beef breeds. They do well, however, on poorer feed than beef cattle and need less care than dairy cattle.

Below is a technical explanation consisting totally of a classification of drill accessories based on functional standards. To help make the classification easy to understand, the writer inserted graphic illustrations.

> A drill bit has a working end that makes holes and a smooth shank that is grasped by the jaws of a chuck. While bits can be bought individually, they cost less if purchased in sets.
>
> The twist bit (Figure 6.2), the most commonly used, cuts cylindrical holes. It has a sharp point and two spiral-shaped cutting edges that lift chips out of the hole as the bit turns.

Figure 6.2
Twist bit.

Carbon steel twist bits are suited to drilling wood and soft metals; high-speed steel bits cut wood, soft metals, and mild steel; tungsten carbide or carbide-tipped bits cut hard metals and masonry. Cutting diameters commonly available range from one-sixteenth to one-half inch.

The wood-screw pilot bit (Figure 6.3) has three widths of cutting edge. The least wide drills a hole to give screw threads solid anchorage. The next makes a shaft for the unthreaded screw shank. The widest makes a hole, or countersink, for flatheaded screws. A detachable stop can make shallow or deep countersinks.

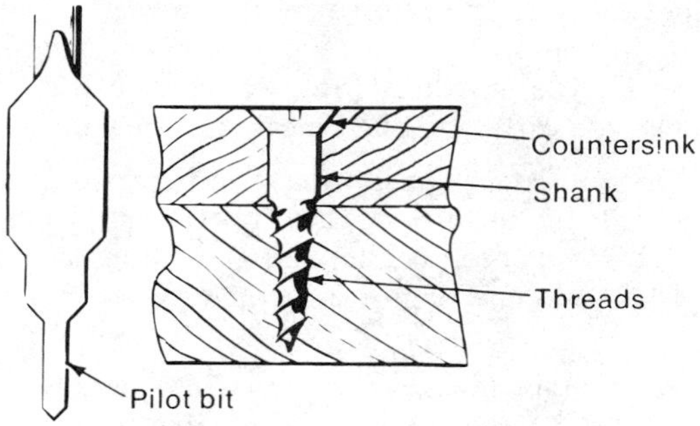

***Figure* 6.3**
Wood-screw bit.

The screw-driving bit (Figure 6.4) attaches to drills with variable speed and reverse to drive and remove slotted and Phillips-head screws. On single- or two-speed drills, the bit must be used with a screw-driving attachment.

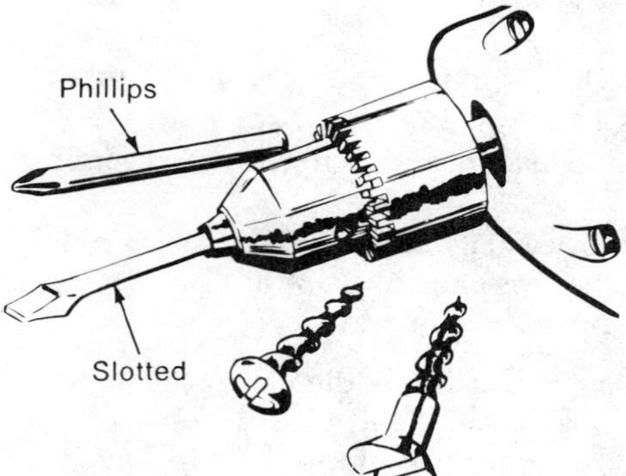

***Figure* 6.4**
The screw-driving bit.[2]

[2] United States Navy, *Tools and Their Uses* (Washington, D.C.: U.S. Government Printing Office, 1971), p. 53.

DEFINITION

Most speaking or writing about technical subjects requires some use of definition. Whenever you tell what something is or what it means, you are defining. You can't succeed in informing or persuading someone unless you both agree on what your words and ideas mean. Anytime you even suspect that your receivers might not understand something as you want them to, you should define it.

A writer or speaker may use definition either as a total communication or merely as a supporting section. When you devote the complete communication to telling what the legal term *caveat emptor* means, for example, the total communication is a definition.

When definition is used in technical communication as a supporting form, it may be any of the following most common kinds: (1) basic, (2) synonym, (3) extended.

Basic Definition

The most often used form of definition is basic definition. Whenever you are asked what something is or what it means, you automatically reply with a basic definition. For example: "What is a bugle?" "It is a brass wind instrument without keys."

Basic definitions consist of three main parts:

1. The *term*—the word or words being defined.
2. *Classification*—the class to which the item represented by the term belongs.
3. *Differentiation*—how the word or idea is different from the others in its class.

Following are examples of basic definition:

Term	*Class*	*Differentiation*
lapidary	a person	who cuts, polishes, or engraves gems
keystone	a central wedge-shaped stone in an arch	that locks the other stones together
digitalis	a drug	used as a heart stimulant

Three of the most common kinds of definition differentiation are:

1. *Structural*—points out differences in the parts or the overall structure of the term defined from the other items in its class.
2. *Functional*—tells what it does that is different from the others.
3. *Causal*—states the differences between its causes and the others.

Here are some examples of these kinds of differentiation:

Term	*Class*	*Differentiation*
		Structural
scissors	cutting tool	consisting of two blades each with a loop handle and joined by a swivel pin.
tripod	stand	with three adjustable legs.
		Functional
ferrule	metal ring or cap	used to reinforce the end of a wooden pole, cane, or handle.
ferret	narrow tape	used to bind or edge fabric.
		Causal
toggle joint	elbowlike joint	with two arms pivoted so that a force applied to the hinge causes them to straighten and produce an outward force at their ends.
diabetes	metabolic disorder	caused by insulin deficiency.

Synonym Definition

The easiest kind of definition is a synonym. It is just another word or group of words for the term defined. Since it means the same thing as the defined word, it is often used when the writer thinks the reader may already understand what is meant, but just wants to be sure. A synonym also enables the speaker or writer to move receivers of the message from what they understand to something they don't understand as well, or the other way around. Doing this will help you to inform or to persuade your listeners or readers more effectively.

Learn to use synonyms to give definitions in a flash, so that you don't distract your receivers from your main ideas. Here are some examples with the synonym definition italicized:

> Tsunamis, *tidal waves,* caused fewer than 1,500 deaths between 1900 and 1968.
>
> A burin, a *cutting tool,* was in the engraver's hand.

Notice how the synonyms in the following supporting paragraph help define quickly, easily, and concisely, to avoid distracting the receiver:

> During the disappearance of the glaciers through melting, unsorted mixtures of clay, sand, gravel, and boulders, *till,* were generally deposited by the ice sheets as an unconsolidated mantle on the countryside, *ground moraine.* In some places, hummocky belts, *narrow ridges,* may occur, or drumlins, *swarms of rounded elongated hills.*

Extended Definition

Extended definition is simply an expansion or amplification of the differentiation of a basic definition. Since differentiations are structural, functional, or causal in nature, the extended definition will usually enlarge upon the term defined in a basic definition by telling what it looks like, how it functions, or why it looks or functions as it does. Sometimes a combination of these kinds of differentiation is used. A section of definition, consisting of more than one supporting paragraph, is not uncommon.

The differentiation of the following basic definition is extended mainly by using structural enumeration.

> The jack hammer, a necessity on many construction jobs, is a hand-operated tool used to cut through hard surfaces. It weighs from sixty to ninety pounds. It measures about two and one half feet in height. Its double handle runs parallel to the ground when the hammer is in the upright position. The on/off control lever is attached to the right handle. A variety of cutting attachments may be fastened to its point, depending upon the nature of the surface and what is to be done to it. The three most often used attachments are the point, the wedge, and the spade.

The next basic definition is extended mainly by functional enumeration. It mainly tells the functions its components perform.

> "Scuba" is a brief form for "self-contained underwater breathing apparatus." The scuba itself consists of the cylinder, a cylinder valve, and a regulator. The cylinder is used to hold a supply of air under intense pressure. The cylinder valve controls the air supply to the regulator. The regulator is the backbone of the apparatus. It adjusts the air pressure in the cylinder to the pressure of the area surrounding it.

The differentiation of the following basic definition is developed and extended by causal analysis. It mainly explains the reasons why a battery creates and stores electrical energy.

> Essentially a battery is a large sturdy tank that generates an electrical current. Its interior is divided into a number of cells. Each cell contains two sets of lead plates submerged in a solution of sulphuric acid. When the cell is working, a chemical reaction takes place between the lead and acid. This reaction creates positive and negative electrical charges on the plates. An electrical current (two volts per cell) flows from the negative charged plates to the ones with a positive charge. Connecting the plates in each cell causes the combined electrical current to flow to the positive terminal, where it can be tapped for use.

Application 6–1

Examine Figure 6.5 and list three standards by which these tools can be classified. The first should be structural; the second, functional; and the third, causal. Also, give three general class names into which all of them can be sorted.

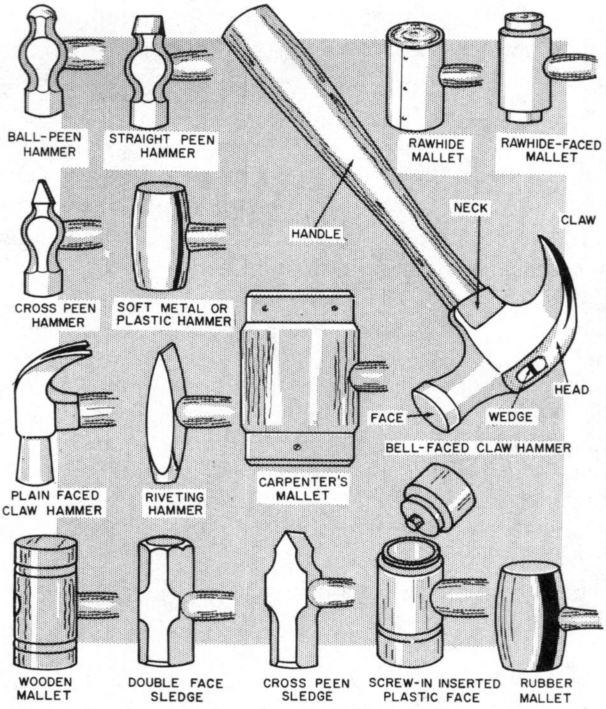

Figure 6.5
Tools for pounding.[3]

Application 6–2 Using your answers for the preceding application, write at least three separate supporting paragraphs, at least one for each of the standards listed above.

Application 6–3 Write four separate supporting paragraphs. Each paragraph should develop a classification of "groceries" based on each of the following standards. Be sure that the term to be classified and the standards by which the classification is to be made are identified:

1. Packaging: cans, glass, plastic, cellophane.
2. Preparing and serving: frying, boiling, baking, roasting.
3. Structure: size, color, texture, nutrition content, etc.
4. Function: how it affects health, weight, strength, etc.

Application 6–4 Write a paragraph of classification that sorts out animals, insects, fish, or birds by each of the following kinds of standards:

1. Structural standards.
2. Functional standards.
3. Causal standards.

[3] U.S. Navy, *Tools and Their Uses,* p. 3.

Application 6–5 Select any two of the following and write two compositions. The first should be developed by structural classification and the second by functional classification. Be sure to indicate which is functional and which is structural.

Shoes	Students
Food	Trucks
Notebooks	Kitchen utensils
Crime	Luggage

Application 6–6 The following sentences call for development by classification. Select two of them and write a supporting paragraph for each.

1. Any metropolis is a composite of four main types of buildings.
2. College students often exhibit a variety of attitudes toward scholarships.
3. Every zoo usually has several types of carnivorous animals.
4. There are three main causes for juvenile delinquency.
5. The houses in the development fall into one of five architectural designs.

Application 6–7 Examine Figure 6.6 and write three kinds of basic definition: (1) structural, (2) functional, (3) causal. Above each of the definitions identify the term, the class, the differentiation, and the kind of differentiation intended.

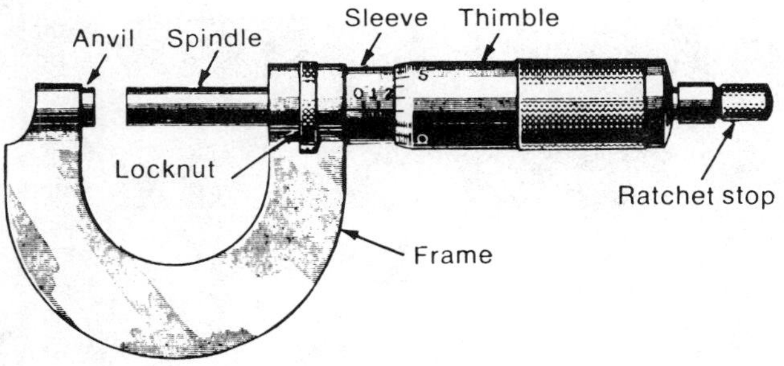

***Figure** 6.6*
Micrometer caliper.[4]

Application 6–8 Extend each of the definitions in the preceding application into a supporting paragraph, consisting of at least four sentences.

Application 6–9 Print an *F* before each definition below that contains a functional differentiation; print *S* before each that uses a structural differentiation or print C before each that uses causal differentiation. Also, circle each term defined.

_____ 1. Japan is a resin varnish mixed in paints to act as a drying ingredient.

_____ 2. An electron is a particle of matter with a negative charge.

[4] U.S. Navy, *Tools and Their Uses*, p. 18.

_____ 3. Enzymes are organic substances that cause changes in other substances.

_____ 4. A drill-press is a machine tool used for producing holes in metal.

_____ 5. An ennead is a group or set of nine.

_____ 6. An envoy is an agent sent by a government or ruler to transact diplomatic business.

_____ 7. A duckbill is a small water mammal with webbed feet, a tail like a beaver's, and a bill like a duck's.

_____ 8. A dry battery is an electric battery made up of several connected dry cells.

_____ 9. An entablature is a horizontal superstructure supported by columns and composed of architrave, frieze, and cornice.

_____10. A dry cell is a voltaic cell that is either sealed or treated with an absorbent so that its contents cannot spill.

Application 6–10

Examine the pictogram below, which shows several categories of school dropouts. From the facts in it and the inferences you draw from it, write a supporting section of at least three paragraphs developed by classification and definition.

850,000
Elementary & secondary school dropouts; many found school irrelevant
750,000
General curriculum high school graduates who did not attend college
850,000
High school students who entered college in 1967, but did not complete the baccalaureate or an organized occupational program
TOTAL 2,450,000 (est.)

SOURCE: U.S. Department of Health, Education, and Welfare, Office of Education, "Career Education" (Washington, D.C.: U.S. Government Printing Office, 1971), p. 9.

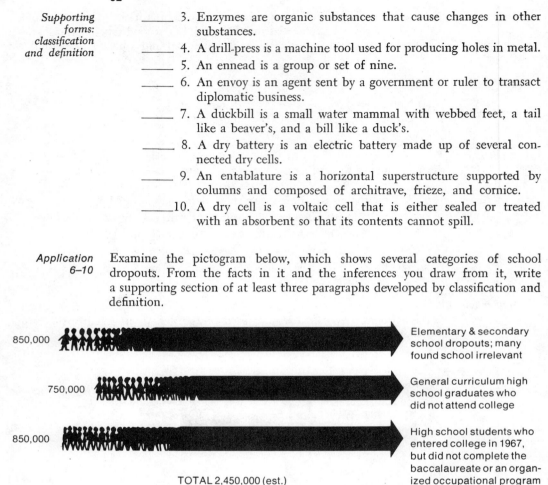

7

Supporting forms: illustration, comparison, contrast, analogy

what they are . . . how they differ in form and application . . . their purposes . . . using them skillfully

In this chapter we will study four more supporting forms—illustration. comparison, contrast, and analogy. These patterns of development are especially useful in amplifying description, narration, and causation. Using them effectively also enables a speaker or writer to adapt technical ideas to the background and interests of intended readers to inform or persuade them.

Below are brief definitions for these four often used supporting forms:

Illustration is offering the reader an example, a sample—in a sense, a specimen of something.

Comparison is pointing out the similarities between items belonging to the same class.

Contrast is pointing out the differences between items belonging to the same class.

Analogy is pointing out the similarities or differences between items not belonging to the same class.

ILLUSTRATION

One of the easiest ways to enable the intended receivers of your messages to relate new ideas to something they already understand is by illustration. The illustration used may be a single word or a series of ideas that make your meaning clear or that act as evidence in support of a conclusion.

Following is an example of illustration used in expanding a definition and clarifying it by relating it to something with which the reader is more familiar:

> Magnetoelectricity is produced when a good conductor, copper, for example, is passed through a magnetic field and the force of the field causes the conductor atoms to free their electrons. These electrons will then move in a certain direction.

Below is another good example of the way you may resort to illustration to adapt your ideas to the background and interests of your intended receiver. You can safely assume that the other person is familiar with "cables on a crane," the "foundation of a building," and the way a pair of scissors cuts cloth to understand the different kinds of stress:

> Tensile, compression, and shearing are three fundamental stresses. Tensile stress, or tension, is caused by a force that tends to pull-out or stretch a piece of material. The cables on a crane are subjected to tensile stress everytime they lift a load. Compression, or compressive stress, is caused by a force tending to press together, to shorten a material. The foundation of a building or the base of a heavy machine is undergoing continuous compressive stress. Shear, or shearing stress, is caused by forces that cause one part of a piece of material to slide over another part of it, as a pair of scissors does to cut cloth. Punching holes in boiler plate and cutting sheet metal requires shear stress. These stresses are extensively applied in industry and engineering.

A writer must be able to judge how much illustration and other forms of explanation the reader needs to understand what is being said. The illustrations need not be limited to only one supporting paragraph. Often, several paragraphs are used to illustrate a point. In narrative illustration, it wouldn't be unusual for the writer to use a section of several paragraphs to make a point clear.

> If you look at different types of vending businesses, you'll see that the operators started with varying kinds of experience. Not all of them had the same background when they went into vending.

Illustration

Take Virgil Lund. He had an average job with an accounting firm in a town of 126,000. Then his father-in-law suggested that he try his hand in the family vending business. Lund had no vending experience, but his business background and training in accounting came in handy. He didn't exactly start at the bottom; and now that his father-in-law has retired, he is the manager. But he still goes out on service calls.

Jack Peebler, on the other hand, was a service-station mechanic in a suburb of Boston. Several times, he had used his mechanical skills when the soft-drink machine where he worked was out of order.

He thought he saw an opportunity in the many small offices and factories in his neighborhood and inquired about how the managers were taking care of the coffee breaks. With some financing and other help from a vending machine manufacturer, he placed 12 instant-coffee machines, servicing them in his spare time. Because he was good with machinery, he could handle service and repair problems in the evenings and on weekends. Accounting and other recordkeeping were harder to manage; but his wife had taken a business-college course, and she took charge of the bookkeeping.

Within a few months Peebler saw possibilities for expansion. He quit his job and now works at his vending business full time.[1]

COMPARISON

Comparison is a common, easy-to-use form of developing supporting paragraphs to inform or persuade. It is especially useful in developing description or narration by pointing out the similarities between subjects. Below is an example of description developed by comparison:

> Thomas Jefferson and Alexander Hamilton were alike in their political attitudes. Neither man was an extremist. Both believed in a republic as the best form of government. Both believed in democratic principles, but they also favored putting the political power in the hands of men with education and property. Both realized that state powers and state interests could not be ignored. They also favored the creation of a military force, including a strong navy. Each was a fair man, playing the game of politics within the rules that democracy must have.

The comparison of Jefferson and Hamilton in the preceding illustration is developed by enumeration. When it will serve the purpose, enumeration is a useful way to develop any of the supporting forms. However, when

---

[1] Small Business Administration, Office of Management Assistance, *Starting and Managing a Small Business* (Washington, D.C.: U.S. Government Printing Office, 1967), p. 9.

a more detailed treatment is needed, you may have to develop your comparison by analysis. Following is an example of a comparison developed by analysis. Notice how the analysis explains the relationships between each step in the process of reproduction.

> The common characteristics that whisk fern and liverwort share is the habit of reproducing by means of sexual generation that is dependent upon water. The whisk fern releases tiny spores from the cases on its branches. These spores grow into tiny plants and live underground, covered with male and female sex organs. In the damp earth, fertilization takes place by the sperms' moving over to the eggs. After fertilization, a plant shoot develops and another spore-producing plant emerges. Like the fern, liverworts cluster in damp places. The male produces sperm in the disks, and rain water spreads it in the female plants. Shortly after fertilization, another plant emerges. By comparing the whisk fern and the liverwort, two early forms of lichens, we can understand better their dependence upon water for reproduction.

Did you notice that the comparisons in the preceding illustrations are between two persons and objects belonging to the same class? Consequently, the comparisons are logical. It would be illogical to develop a comparison between the eating or reproduction cycles of antelope and those of salmon because antelope and salmon do not belong to the same class. For the same reason, it is not logical to develop a supporting paragraph of contrast between them.

All of the supporting forms, including comparison, contrast, and analogy, need not be limited to a single paragraph. You may find it necessary to use a section, consisting of a series of paragraphs, to develop any of them.

CONTRAST

Contrast is pointing out the differences between things belonging to the same class. Since they belong to the same class, the items being contrasted must have some traits in common, but the speaker or writer is mainly concerned with their differences.

Contrast has more impact than comparison because it expresses ideas in sharp relief, like placing white chalk marks on a blackboard. You can see how the following contrast emphasizes the performance and safety advantages of radial tires.

> The radial tire is more durable than the conventional tire because it is designed and constructed for performance and safety. The design of the radial is different; it always looks underinflated. This design is intended to make it safer to drive by enabling better road traction. The regular tire, on the other hand, always looks well filled with air, and although it provides a smoother ride, it doesn't hold the road as well. Unlike

conventional tires, radial tires are built with steel cords rather than cords of nylon or fiberglass, resulting in longer life. In radial tires the cords are wrapped around the casing in a parallel direction, giving additional strength. Conventional tires are not wrapped at an angle and do not hold up as well. Because of their additional strength and safety features, radial tires are being installed as part of the standard equipment on many new automobiles.

When you are sure that the receiver of your message is familiar with one of the terms in your comparison or contrast, you only may have to discuss one of the terms. Here is an example of a contrast developed by focusing attention on the "first automobiles." The writer knows that the reader will relate the characteristics pointed out to those of today's cars.

> Although the first automobiles traveled at slower speeds in less traffic, they were inconvenient, uncomfortable, and more dangerous to drive than today's automobiles. In the early 1900's the "horseless carriage" came into being. Motoring in those days was something of an adventure. Drivers had to wear goggles, for there were no windshields. Everyone wore linen dusters to protect clothing against the clouds of smoke. Early motorists had to carry their own gasoline because there were no filling stations. Tire chains were a necessity every time it rained. Towropes had to be used each time the car got bogged down in a mudhole. Blowouts had to be repaired with a patching kit as there were no removable spare wheels.

As stated earlier in our dicsussion of comparison, a supporting section of contrast does not have to be limited to a single paragraph. Sometimes a series of paragraphs may be needed to develop contrast enough for the reader. The following series of light, fluffy paragraphs is intended for readers just getting acquainted with tent and travel trailers.

> There are several significant differences between the tent and travel trailers that a potential buyer should consider. The type of traveling you are going to do will be a big factor in your choice. If you plan to go to Van Ettan Lake in Oscoda, Michigan, and stay for several days or weeks in the late spring, summer, or early fall, a tent trailer is an economical and adequate mode of camping. On the other hand if you plan to travel all day, arriving at the campground late, and don't want all the fuss of setting up the tent trailer, pulling out the gear just so you can get into the rig, then the solid-side travel trailer is for you. All you need to do in this type of rig is unlock the door. The rest is set up.
>
> Another item that is big on the "what to look for" list is the bathroom. Most travel trailers have one complete unit with shower and sometimes a bathtub. The tent trailers very rarely have any kind of bathroom—so with a tent trailer you are forced to seek a campsite close to restroom facilities. This is an especially significant item if there are small children in the family.

The tent trailer, because of its construction, usually weighs less than its counterpart of comparable length. The license fee, if by weight, comes into play here. A Rover Convertible 8 sleeper tent trailer 1,000 pounds weight, cost $4.00 to license. A Pirate 6 sleeper travel trailer, weighing 3,110 pounds, costs $12.50 to license. Towing a folded-down tent trailer at 50 MPH decreases the mileage by one to two MPG. Towing the travel trailer at the same speed will cost between 4 MPG and 7 MPG depending on terrain. The entire problem boils down to "What do I want?" Do I want luxury and comfort at the sacrifice of economy, or do I want economy and the more rugged life of a tent trailer?

ANALOGY

Analogy is developed by pointing out or explaining the likenesses or differences between things not belonging to the same class. Therefore, the comparison or contrast in an analogy may not be logical except in a special way.

The writer of the following paragraph enables readers to understand the function of a shock absorber by drawing an analogy. Notice how there is only a special kind of similarity between the shock absorber and the rubber ball.

> The purpose of shock absorbers on an automobile is the arresting of excessive movement of the springs on both the up and down thrusts. By preventing excessive rebound, the shocks keep all four wheels in firm contact with the road under virtually all driving conditions.
>
> The dampening function of a shock absorber on the spring action and ride control is much like putting liquid in an inflated rubber ball. A rubber ball rolled over a rough surface would normally bounce out of control on each contact with a bump or depression. Adding liquid would control these radical fluctuations. In much the same way shock absorbers reduce the adverse effect on car control brought on by bad roads, hard braking, and high-speed cornering.

Although analogy is often used in fiction and poetry to make an abstract idea concrete and vivid, it is just as often used in commercial and technical writing for the same reason. Notice how analogy in the following section enables us to understand much more clearly the principle discussed:

> Living organisms continuously interact with their environment. They must constantly take in nutrients and water and return waste products to the environment. This is analogous to what Heraclitus, an ancient Greek, taught. Our bodies, he said, are like rivulets, their material renewed like water in a stream.

Analogy

> To illustrate this concept, Oparin, a Russian biochemist, cited the following analogy. A bucket of water is a closed system. If evaporation is prevented, the water remains at a constant level. But a tank of running water, with an inlet and outlet pipe, also remains at a constant level, provided inflow exactly equals outflow. By changing the rate of either inflow or outflow temporarily, a new level can be established. In the sense of inflow and outflow, such a tank can be compared to a living organism with its intake of food and water and discharge of waste products to the environment.[2]

In this final example of an analogy, the comparison (in italics), if taken literally, would be illogical.

> A sprinter running the hundred-yard dash is like a *well-oiled piece of machinery*. The bang from the gun is the "on" switch that starts the runner moving. He is propelled down the lane by the pistonlike movement of his arms and legs. This movement isn't jerky. It is smooth and rhythmic with no motion wasted from side-to-side movement. Every part moves up, down, up, down, up, down, except the head, which is held steady with eyes fixed on the finish line. You don't have to use your mind for strategy planning in this race. It is an all-out burst of energy with speed being the most important aspect. When the sprinter feels the tape cross his chest it's as if he is gradually turned off. He slows and finally stops. The machinelike quality is lost and the sprinter becomes a normal human being after a race, breathing unevenly and walking with slow, stumbling steps.

It should be emphasized here that all of the supporting forms may be used in various combinations. Definition often needs the help of illustration to adapt it to the reader's background. Comparison and contrast are often combined. Here is a supporting paragraph developed by a combination of comparison, contrast, and illustration.

> The raising of buffalos in large herds just as cattle are raised is increasing on large ranches. Like cattle hides, buffalo skins are very suitable for the manufacture of shoes. Jackets, vests, and a variety of other kinds of clothing now made of cow hides may be made from that of buffalos. But much of this clothing can be made at less cost because buffalos are easier and cheaper to raise. They do not need as much care as cattle, and during the winter they care for themselves. For example, they will paw through the snow to find feed, but cattle will starve to death if you don't provide the feed. During calving, buffalos will take care of themselves. For these reasons, farmers in Indiana, Ohio, and Michigan are increasing the size of their buffalo herds.

---

[2] Eugenia Keller, "The Origin of Life," *Chemistry*, December 1968, p. 11.

Application 7–1 Develop a description of the column capitals pictured in Figure 7.1. Begin with a comparison, then write a supporting section of contrast.

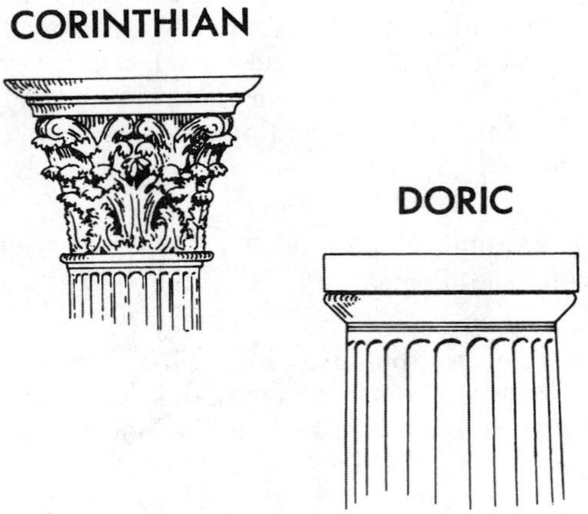

Figure 7.1
Column capitals.

Application 7–2 Write a section of causation or description expanded by analogy to enable a layman to understand any construction features of the bridges illustrated in Figure 7.2.

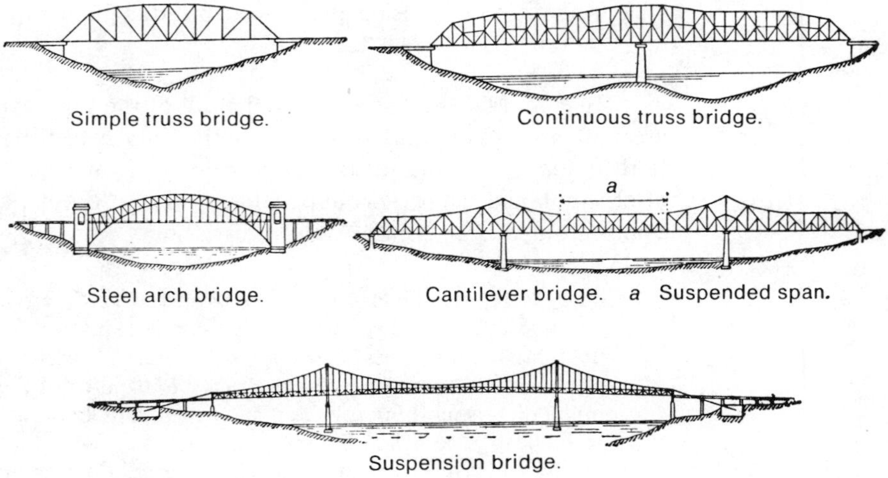

Figure 7.2
Bridge construction.

Application 7–3 Examine the pictures of the special-purpose saws shown in Figure 7.3, and write a supporting section of description, narration, or causation, containing comparison and contrast. Also, use illustration to expand the comparison or contrast in explaining its use.

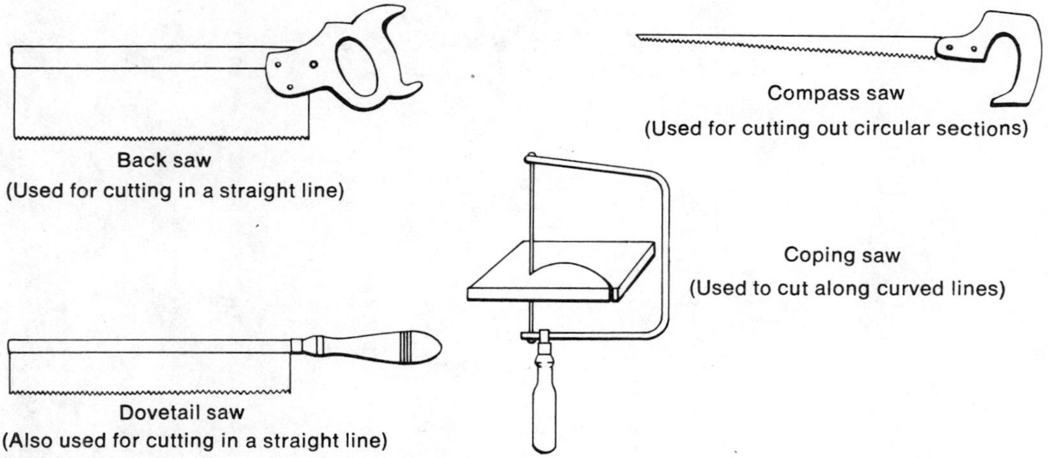

Figure 7.3
Special-purpose saws.

Application 7–4 Select Table 7.1 or 7.2 and write a supporting section of comparison or contrast about it. Be sure to use specific illustration to clarify costs.

Table 7.1 Comparative college expenses, 1948 and 1968

	1948	*1968*
Tuition and fees, by semesters	$300	$720
Books, per year	40	76
Room, semester contract	75	240
Board per six weeks period (6 periods)	36	108
Clothes, per year	80	140
Spending money per week (36 weeks)	5	8

Table 7.2 Percentage distribution of wholesale food sales, by departments, 1972 and 1973

Department	1972 *(average, 20 firms)*	1973 *(average, 19 firms)*	1972–73 *average*
	Percent	*Percent*	*Percent*
Grocery	72.8	70.3	71.5
Meat	14.5	17.5	16.0
Produce	5.6	7.0	6.3
Frozen food	7.1	5.2	6.2
	Million dollars	Million dollars	Million dollars
Total sales	490	646	568

Table 7.3 Changes Occurring in the Major Fleets of the World by Number of Ships During 1965

	Additions				*Deletions*					
Country of registry	*New construction*	*Transfers in*	*Other additions*	*Total*	*Transfers out*	*Losses*	*Scrappings*	*Other deletions*	*Total*	*Net change*
France	9	4	1	14	36	1	3	4	44	− 30
Germany (West)	21	18	3	42	26	3	9	—	38	+ 4
Greece	9	135	—	144	63	11	16	3	93	+ 51
Italy	15	19	2	36	13	4	22	1	40	− 4
Japan	74	1	23	98	10	3	23	2	38	+ 60
Liberia	74	165	1	240	41	12	28	—	81	+159
Netherlands	7	2	—	9	23	2	4	—	29	− 20
Norway	86	11	8	105	97	3	24	2	126	− 21
Panama	4	67	2	73	25	8	25	3	61	+ 12
U.S.S.R.	109	—	16	125	2	—	1	4	7	+118
United Kingdom	76	13	3	92	103	3	26	5	137	− 45
United States	13	1	15	29	13	1	135	42	191	−162
Total	497	436	74	1,007	452	51	316	66	885	+122

SOURCE: U.S. Department of Commerce, Office of Maritime Promotion, "Merchant Fleets of the World" (Washington, D.C.: U.S. Government Printing Office, December 31, 1966), p. 11.

Application 7–5 Study Table 7.3. Find the totals which are approximately the same under each column and write a supporting section consisting of at least three paragraphs of comparison.

Application 7–6 Study the bar graph in Figure 7.4 and write one supporting paragraph of comparison and another of contrast.

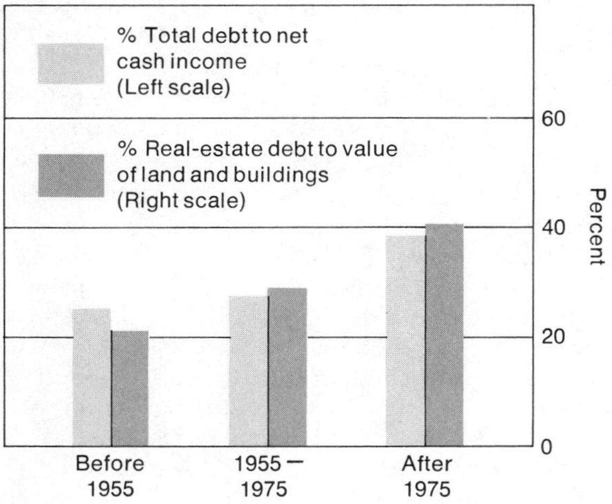

Figure 7.4
Relative indebtedness of major real-estate borrowers.

Application 7–7 Select one of the sets of words in the following list and write a supporting paragraph of comparison and another of contrast.

good and bad advertising
college and high school
service in the army, navy, and marines
two tools, apparatus, or ingredients
two students or employees

Application 7–8 Select one of the sets of words below and write an analogy to clarify the first word in the set.

a human eye—a camera
a heart—a pump
a brain—a computer
radar—rubber bullets
communication—ping pong
slang—firecrackers

Application 7–9 Each of the following supporting paragraphs is developed by illustration. Underline the illustrations.

Supporting forms: illustration, contrast, analogy

1. There are times when, for engineering purposes, a definite force must be applied to a nut or bolt head. In such cases, a torque wrench must be used. For example, equal force must be applied to all head bolts of an engine. Otherwise, one bolt may bear the brunt of the force of internal combustion and ultimately cause engine failure.
2. A one-way design fabric such as a print or a weave has a pattern that moves in one direction only—for example, a flower print with the flower heads pointing the same way. When you deal with this type of fabric, when you lay out your pattern pieces, all the pieces must have their tops pointing in the same direction so that the design will not be interrupted where the pieces are sewn together. Stripes and plaids must be treated in the same manner. Solid colors or designs require less material because matching pieces are not necessary for construction.
3. Simple precautions can prevent many of the automobile accidents that occur on our congested city streets. One major cause of these senseless accidents is motorists' concentrating on everything but driving—Mr. Smith's thinking over the problems he has to solve at the office, Mrs. Jones' concentrating on what she'll fix for dinner tonight, and Sally's dreaming about that date she has planned. Because of his lack of concentration, the motorist may not see the car in front of him slow down, or the small child wandering into the street. By trying to concentrate on what's happening around him, checking the rear view mirror, or slowing down when approaching parked cars, the driver can avoid a serious accident. A second cause of these unnecessary accidents is the attitude of some drivers that the other "guy" is responsible for his own safety. To Mr. Jones, dashing out from a side street directly in front of an oncoming car, it isn't his fault that another car had to swerve into the other lane, causing an accident. By trying to realize that the other person can't read minds, by driving defensively, drivers can prevent many accidents.

Application 7–10

Select one of the items listed below, and write a supporting paragraph or two using illustration to clarify a point or to act as evidence in support of your conclusion. Tell whether your purpose in writing the paragraph is to inform or to persuade your reader, and indicate the purpose for which your illustration is intended. Underline your illustrations.

rafters
A-frames
hats
wigs
television commercials
frozen foods
planes
kitchen ranges
automatic dishwashers
restaurants

8

The nonverbal message—graphic illustrations

common kinds of graphic illustration: tables, graphs, drawings . . . what they do . . . preparing them . . . using them skillfully

Technical writers often use devices other than words to help express their ideas clearly. They rely on various kinds of visual devices to make complex relationships between their thoughts easier to understand. By telling and showing at the same time, they enable technically oriented people to interpret data at a glance and to arrive at difficult conclusions quickly. Following are some of the more common general kinds of graphic illustrations you will find helpful:

1. Tables.
2. Graphs.
3. Drawings.

Tables, graphs, and drawings mainly are nonverbal, pictorial forms of illustration. They are used to amplify words and sentences so that these verbal elements do a better job of developing description, narration, and causation. They enable readers to see the structural, functional, and causal relationships that the words express and, consequently, to understand them better.

The nonverbal message—graphic illustrations

Graphics are also used to make clear the relationships between the details in the supporting forms used to develop description, narration, and causation. They are often used to enable readers to understand better by seeing the similarities or differences in comparison, contrast, analogy, definition, classification, and illustration.

Graphics are word helpers; do not use them instead of words. Use them when the words alone cannot adequately inform or persuade the type of reader at whom you are aiming. Because they express complex ideas clearly and accurately, graphics may help you to reduce the number of words needed to explain a complex idea.

Place a graphic as close as possible to its reference in the text, preferably on the same page. A table, graph, or drawing should not precede its mention in the text, unless that is unavoidable.

TABLES

A table is a kind of graphic illustration presenting informative details arranged in rows and columns. By using tables, you will enable your readers to identify and understand interrelationships. Carefully read the following paragraph, and when you think you clearly understand what it says, read the same information in Table 8.1. If you prefer, do it in reverse order; read the table first.

> The following data must be considered before deciding upon the kind of urban transportation required by Fort Wayne, Indiana, and by Fort Worth, Texas. Fort Wayne has 184,002 inhabitants; Fort Worth, 490,-335. Fort Wayne has 55,994 households; Fort Worth, 152,512. Fort Wayne has 59,174 housing units; Fort Worth, 198,977. Fort Wayne has 11 autos per household; Fort Worth, 12.

Table 8.1 Transportation Planning Data

	Fort Wayne	*Fort Worth*
Total population	184,002	490,355
Total no. households	55,994	152,512
Total housing units	59,174	198,977
Autos per household	11	12

Tables may be classified in general as either informal or formal. Informal ones are just lists of details. They enumerate a series of items under column headings. Indexes, tables of content, inventory lists, and specifications are a few kinds of informal tables.

Tables The following table is informal, consisting mainly of a list. Although it has a title, it is not identified with a number.

Vending Machine Sales in July, 1972

	Percent of all sales	*Average sales per machine*
Cigarettes	24.82	$1,496
Candy, nuts, gum	10.92	627
Cold-cup beverages	7.10	1,742
Hot-cup beverages	17.67	2,641
Ice cream	2.10	1,169
Milk	2.72	1,690

Formal tables are designed to present details in a way enabling readers to quickly understand the relationships between them and to arrive at conclusions based on these interrelationships. Unlike informal tables, which do not need to be identified by numbers and titles, formal tables require both.

Table 8.2, a formal table, illustrates how the complex results of an extensive technical investigation can be clearly stated in a small amount of space. The findings summarized are based on coronary heart disease deaths reported over a four-year period among 441,000 men and 563,000 women.

Table 8.2 Coronary heart disease mortality ratios among current cigarette smokers only, by amount smoked daily [1]

		Cigarettes smoked daily			
Age and sex	*Non-smokers*	*Under 10*	*10–19*	*20–39*	*40+*
Men:					
45 to 54	1.0	2.4	3.1	3.1	3.4
55 to 64	1.0	1.5	1.9	2.0	2.1
65 to 74	1.0	1.3	1.6	1.6	*
75 to 84	1.0	1.2	1.4	1.1	—
Women:					
45 to 54	1.0	0.9	2.0	2.7	—
55 to 64	1.0	1.3	1.6	2.0	—
65 to 74	1.0	1.1	1.4	1.9	—
75 to 84	1.0	—	—	—	—

* Expected deaths were less than 10.
Source: E. C. Hammond.

[1] U.S. Department of Health, Education, and Welfare, Public Health Service, *The Health Consequences of Smoking* (Washington, D.C.: U.S. Government Printing Office, 1967), p. 49.

The main sections of a formal table are (examples refer to Table 8.2):

Number—This may be a Roman or an Arabic number preceded by the word "Table." Along with the title, it is placed at the top of the table.

Title—Briefly identifies the contents.

Spanner—A caption that spans over the column heads when there are two or more kinds of details under it. ("Cigarettes smoked daily.")

Column head—A caption that identifies the kind of data in the column under it. ("Under 10, 10–19, 20–39, 40 +.")

Column—The series of details entered vertically under each column head.

Stubhead—A caption that identifies the kind of information in the stubs under it. ("Age and sex.")

Stubs—A caption that identifies the kind of information in each row horizontally opposite it. ("Men, Women.")

Row—The series of details entered horizontally opposite each stub.

Footnotes—Explanatory comments related to one or more of the entries or that identify the source from which the data or the table itself was derived.

Following are some things a technical writer should keep in mind when preparing formal and informal tables:

1. All tables should be made as simple as possible.
2. A table should be placed as close as possible to the part of the prose text making reference to it.
3. Whenever possible, a table should be prepared so that it completely fits on a single page. If more than one page is needed, each overflowing page must be headed with the same title and headings appearing on the first page. The word "continued" should be placed next to the title in parentheses on the succeeding pages.
4. Only the first word and the proper nouns are capitalized in all column headings, stub headings, and stubs.
5. In all headings only clearly understood abbreviations should be used.
6. Stubs should be written consistently, using the same kinds of words or numbers. ("Men—Women," "45 to 54, 55 to 64.") Ditto marks should not be used in place of repeated stubs.
7. An informal table need not be ruled, but a formal table, especially a complex one, should have horizontal and vertical lines to make it easier to read and interpret.
8. A writer may use asterisks, lower-case letters, or Arabic numerals as reference symbols for footnotes, which are placed at the bottom of the table or at the bottom of the page (see Table 8.2).
9. The details of information in a table should be arranged in some order: geographic, chronological, alphabetical, or quantitative.

Application 8–1 Prepare a formal table giving the total enrollment in the college you attend, the distribution of the students into classes (freshman, sophomore, etc.), and their classification as to specialities: liberal arts, education, business, science, etc.

Application 8–2 Prepare an informal table presenting several characteristics or specifications of four different automobiles. Some features you might use are: horsepower, top speed, fuel consumption per mile, price, weight, wheelbase dimensions, etc.

Next, add what you need, and beneath the preceding table prepare a formal one with the same information. Be sure to head it properly.

GRAPHS AND CHARTS

The terms graph and chart are used to identify the same graphic illustrations. Both name figurative devices used to show successive changes in a total amount of something, depending upon what happens to another total. The amount that changes is called the *dependent variable* and the other is the *independent variable*. The independent remains constant even though the amount of the other changes. The dependent variable, however, changes when variations occur in the independent one. The amount of miles your car will travel depends on the amount of gasoline in its tank. The number of miles is the dependent variable and the amount of gasoline in the tank is the independent variable. Another example: the amount of potatoes you are able to buy on a certain day depends upon their cost per pound. The higher their cost, the fewer you get. The bar graph in Figure 8.1 will help you understand the difference between dependent and independent variables.

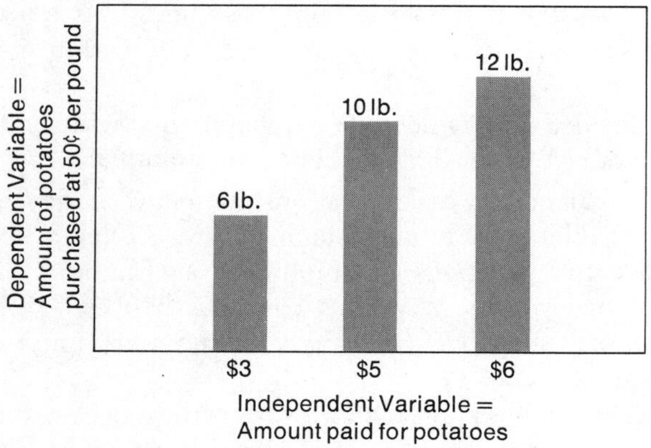

***Figure** 8.1*
Dependent and independent variables.

Following are some of the more common types of graphs or charts:

1. Bar graphs.
2. Pictographs.
3. Pie graphs.
4. Line graphs.
5. Flow charts.

Bar Graphs Bar graphs are often used as visual aids because they are easy to understand and are very effective in immediately capturing the reader's attention. Therefore, they are especially useful when the technical communication is intended for readers who are not experts in a particular field. When writing to experts, the writer would probably use a table rather than a bar graph, because a table is more accurate.

Bar graphs are very effective in showing, along with stating, how two variables contrast with each other. They are highly useful to the technical writer in showing increases or decreases in the total amounts of something in each unit in a series of time periods. If you want your reader to understand the difference in the amount of school-construction expenditures from year to year, you will use a bar graph; for each year involved, you draw a bar of different length to represent the amount spent.

Bar graphs are used most effectively to show amounts that are not continuously changing. To show continuous change in these amounts, you would (as you will see later) select a more effective form of graph, perhaps a line graph.

Depending on the nature of the relationship you want to show, you may select any of several kinds of bar graphs. Each of the graphs in the following list is to some degree more suitable than any of the others for a certain purpose: (1) Basic, (2) Subdivided, (3) Grouped.

Basic Bar Graphs. A basic bar graph consists of a series of two or more separated bars drawn from the same base line. It is especially useful for depicting increase, decrease, growth, or decline within one or more time periods.

Subdivided (or segmented) bar graphs are divided into segments indicating the value of a subvariable. Here is an example:

Basic bar graphs make relationships between dependent and independent variables easier to understand. Figures 8.2 and 8.3 enable the reader to visualize the variations in enrollment at Elmhurst College and the changes in labor costs in various countries between 1974 and 1976. In Figure 8.2 the independent variables are the years, and in Figure 8.3 the country names.

Your bar graphs may be drawn with vertical or horizontal bars. Notice, however, that in the vertical form (Figure 8.2) the dependent variable (the enrollment) is always placed vertically. But in the horizontal form (Figure 8.3) the dependent variable is always horizontal.

Grouped Bar Graphs. Grouped bar graphs are used when all of the bars needed to represent the many variables will not fit on the same graph unless they are grouped. A grouped bar graph consists of three or more bars grouped into a cluster. Each group is drawn with the spaces between the bars omitted. Figure 8.4 is an example of a grouped bar graph.

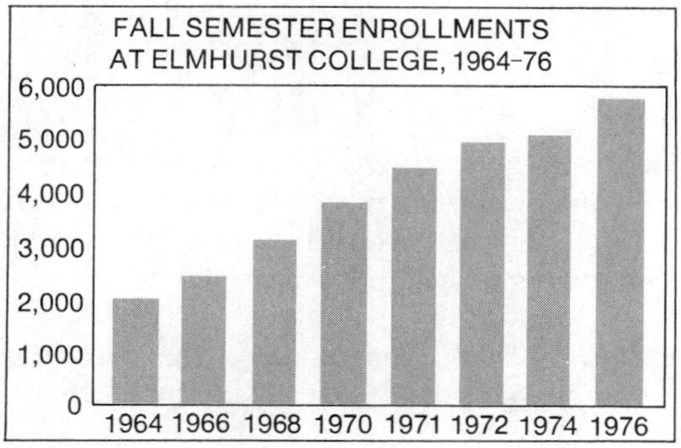

Figure 8.2
Basic vertical bar graph.

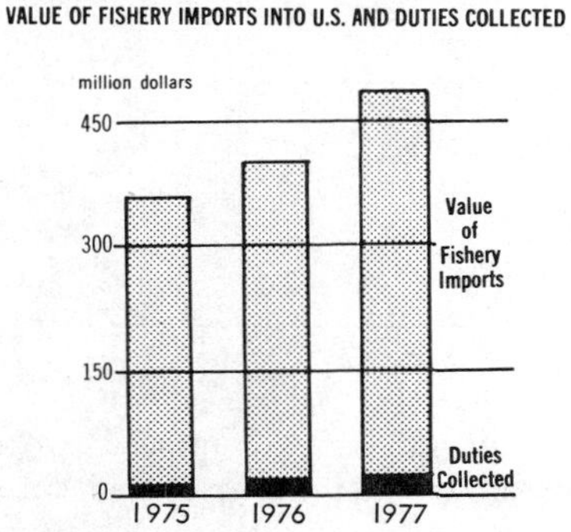

Figure 8.2a
Subdivided bar graph.

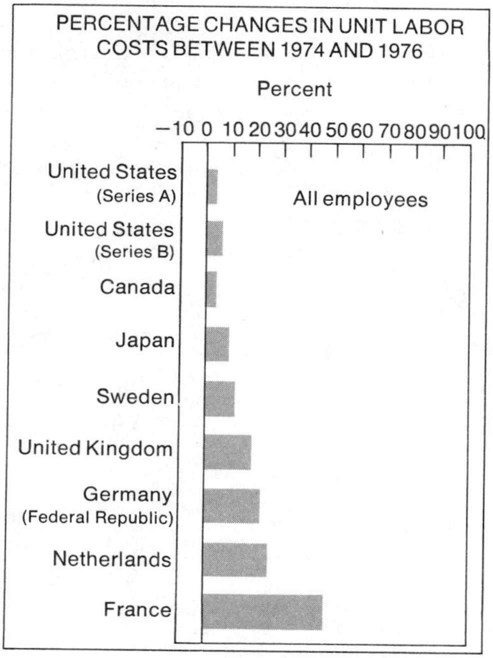

Figure 8.3
Basic horizontal bar graph.

Figure 8.4
Grouped bar graph.

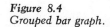

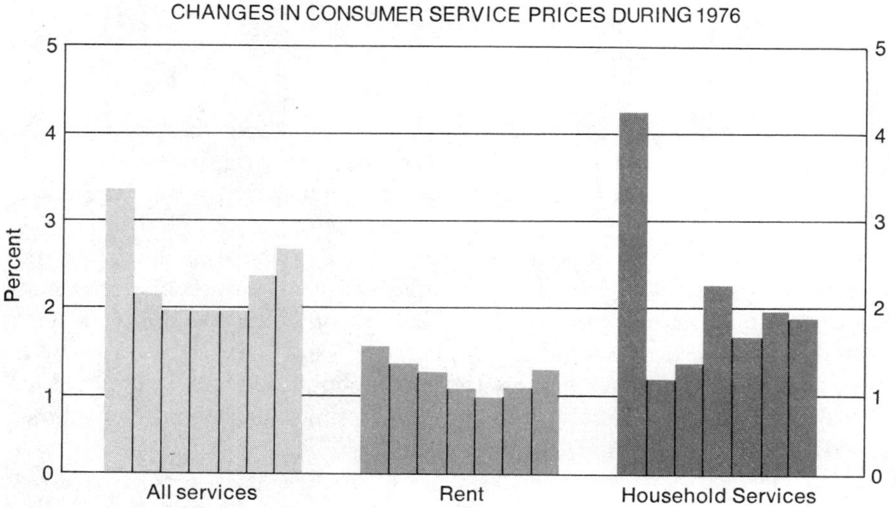

Application 8–3 Make a basic bar graph from the information given in the formal table prepared for Application 8–2. Place a number and title for the graph beneath it.

Application 8–4 Look through various magazines or technical journals from which you may extract examples of the various kinds of bar graphs. Draw your own bar graphs if you cannot find what you need. Now, by means of boxed arrows or arrowheaded lines, identify each of the parts of the graph. Finally, write a short composition that tells what one of the simpler graphs shows.

Application 8–5 Prepare a grouped bar graph that will vividly communicate the following statistics related to the longevity of the employees of a company in your city.

	Percent of employees in various longevity groups		
Year	*Under 10 years*	*10 to 15 years*	*15 to 20 years*
1967	15	35	50
1968	20	42	38
1969	10	51	39
1970	19	40	41
1971	25	35	40

Application 8–6 Between 1967 and 1973, the three main industries employed two million employees. The railroads employed 700,000, water transporation employed 300,-000, and the number of employees in automobile manufacture fluctuated as follows. Prepare a subdivided bar graph for these figures.

1967	750,000
1968	825,000
1969	725,000
1970	915,000
1971	985,000
1972	1,000,244

Draw a grouped bar graph that will point out more emphatically the contrast between employment in railroads and water transportation, which remained constant, and that in the automobile industry, which fluctuated.

Pictographs Instead of bars, a pictograph uses a picture or symbol that looks like the variable it represents. For example, a picture of an automobile represents a certain quantity of automobiles. A drawing of a single soldier in a series of soldiers represents a number who, perhaps, were killed during a battle or a war. When using pictographs, the writer should be sure to indicate the quantity that each picture represents. Sometimes this quantity is shown above, below, or within each image.

Graphs and charts

Figure 8.5 illustrates how vivid the numerical values and the relationships between them become when represented by picture symbols. In the first example, for instance, the pictures of people grow in size to show the population explosion from 49 million in 1964 to 108 million in the year 2000.

WHAT THE POPULATION CHANGE MEANS TO AMERICA

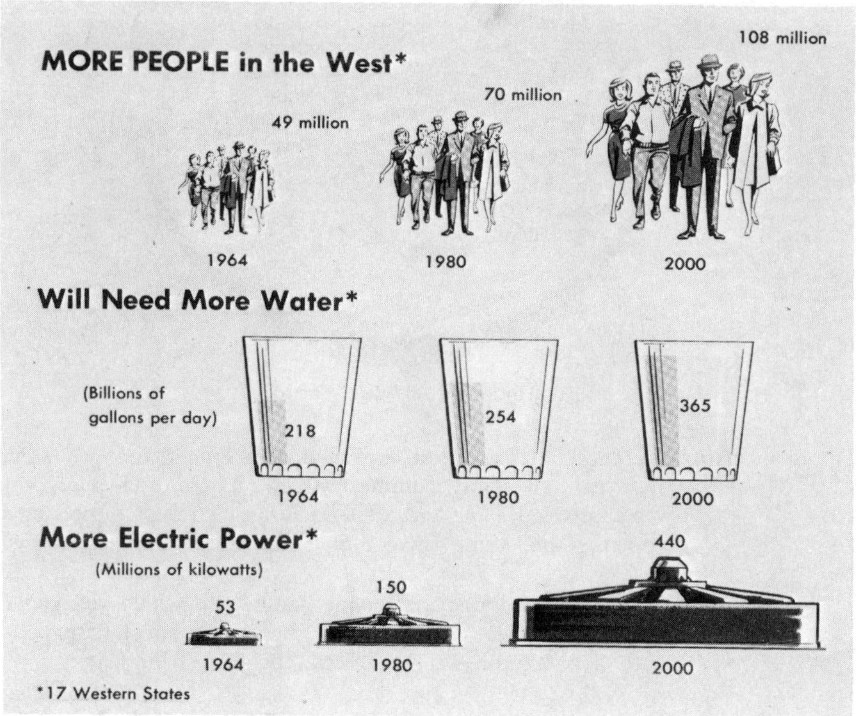

Figure 8.5
A *pictograph.*[2]

Pie Graphs Pie graphs help your reader to see each part of something as a percentage of the total amount. The whole pie represents the total amount, 100 percent. Each slice should be clearly labeled with the percentage relationship it bears to the whole pie.

Notice how vivid and easy to understand the pie graph in Figure 8.6 makes the relationships between the variable amounts and the totals.

---

[2] United States Department of Interior Year Book No. 2 ,*The Population Change —What It Means to America* (Washington, D.C., 1966), p. 25.

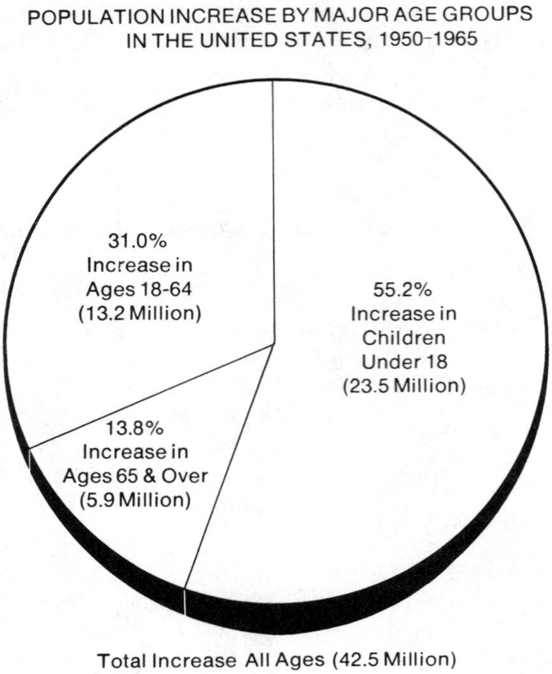

Figure 8.6
A *pie graph*.[3]

Application 8–7 Pretend that you are a hardware store owner and you are preparing a report showing how many hammers, screw drivers, and pliers you sold during the years between 1971 and 1976 inclusive. To depict the totals for each, prepare a pictograph, using images or portions of them.

Application 8–8 Go through a newspaper or magazine from which you can extract pictographs and cut out two samples. Paste these on a sheet of paper and turn them in with a paragraph or two saying the same thing.

Application 8–9 Draw a pie graph showing comparative relationships between the total of your actual or imaginary annual wages and the way your gross income was distributed during a certain year. Be sure to insert the percentage that each slice represents. Place a figure number and title at the bottom of the graph.

Application 8–10 Pretend that you are the owner or manager of a clothing store, grocery, restaurant, or gas station. Using imaginary figures, prepare a pie graph showing what portions of your gross sales were disbursed for labor, supplies, advertising, and maintenance. Be sure to insert in each slice the percentage of the whole that it represents. Be sure to place the figure number and title at the bottom.

Now write a composition of several paragraphs telling in prose what the pie graph shows.

---

[3] U.S. Department of Health, Education, and Welfare, Office of Economic Opportunity, *Dimensions of Poverty in the United States* (Washington, D.C.: U.S. Government Printing Office, 1971), p. 110.

Line Graphs Although not as picturesque or vivid as bar graphs, pie graphs, or pictographs, line graphs are the most useful and accurate in showing continuous change over a period of time.

A line graph is a framework with vertical and horizontal grids (lines) over which one or more curved lines are drawn to show the amount or rate of change occurring in two or more variables.

A variety of kinds of line graphs are used in technical writing; however, this chapter will discuss these two most often used: single-line and multiple-line graphs.

A single-line graph has only one horizontal line, sometimes called a curve. This is drawn between horizontal and vertical scales marked along the frame, which identify the variables plotted by the curve.

Multiple-line graphs have two or more horizontal lines, each representing a related value. If there is enough space, each horizontal line should have a title or an explanation immediately above it to indicate what it represents.

The number of horizontal and vertical grids that either a single- or a multiple-line graph contains depends upon the degree of accuracy required to express the relationships between the variables. When a great degree of accuracy is needed, both the horizontal and vertical grids are drawn completely across the frame. Often, however, as shown in Figures 8.7 and 8.8, the two variables are indicated by grid stubs instead of fully ruled horizontal and vertical grid lines.

Figure 8.7
Single-line graph.

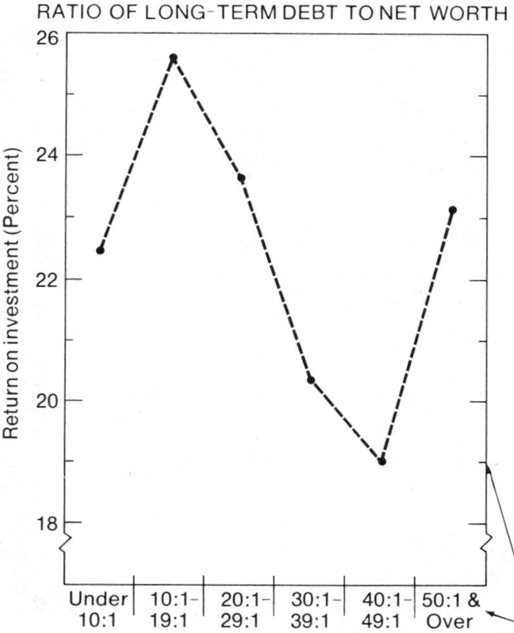

Figure 8.8
Multiple-line graph.

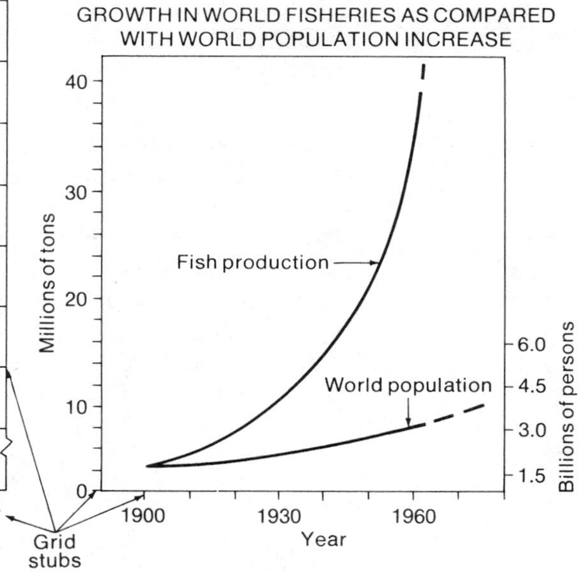

Application 8–11 Look through various magazines for examples of single- and multiple-line graphs. Cut out one of each and paste it on a sheet of paper. Finally, select one of them and write a paragraph or two summarizing what the graph states and the conclusions you draw from it. Tell whether each conclusion you detect is a fact or an inference.

Application 8–12 Plot a single-line graph showing the fluctuations in the amount of rainfall for each day of any month. Next, insert another curve representing the number of automobiles involved in traffic accidents. State any conclusions your multiple-line graph supports. Be sure to place the dependent and the independent variables in the right places. Be sure, also, to give the graph a number and title.

Application 8–13 Prepare a single-line graph plotting the number of miles you drove your car during the past five months. Be sure to identify your graph.

Organizational Charts

Unlike bar graphs and line graphs, charts do not depict numerical relationships between variables. Their main use is to visualize relationships of accountability and responsibility between the divisions of an organization or to conspiculously depict the relationships between the steps or stages of a process or procedure. To objectify these relationships, charts show quantities or values by means of geometric symbols such as rectangles, squares, or circles to indicate the steps or levels. There are two main types: organizational charts and flow charts.

An *organizational chart* is a block diagram showing responsibility functions and the levels of authority within an organization or its units.

The level on which the blocks or other geometric figures are placed in an organizational chart shows that each unit on that line takes directions from divisions represented by the symbols above it and gives directions to those below it. Units having equal authority are placed on the same horizontal level. The flow of authority is indicated by lines drawn between the geometric symbols. The name of the department or division and/or the name of the person who heads each of them are usually placed inside each figure. Notice how the lines of authority and responsibility are clearly depicted in the organizational chart shown in Figure 8.9

Flow Charts

Flow charts are very helpful graphic illustrations. You may use them to make vivid important movement, power, pressure, or steps involved in a procedure. Some are drawn with geometric symbols and others with sketches of the subject. In either type, the sequence or direction is indicated usually by arrow-tipped lines.

In the two illustrations of flow charts, Figure 8.10 contains geometric figures explaining the procedure in "off-line computer control." Figure 8.11 illustrates Newton's third law: when the action and reaction forces are equal in magnitude, a body will remain in unaccelerated movement. The sketch in Figure 8.11 makes it easier for the reader to understand what the forces acting on a plane are and the directions of their effects.

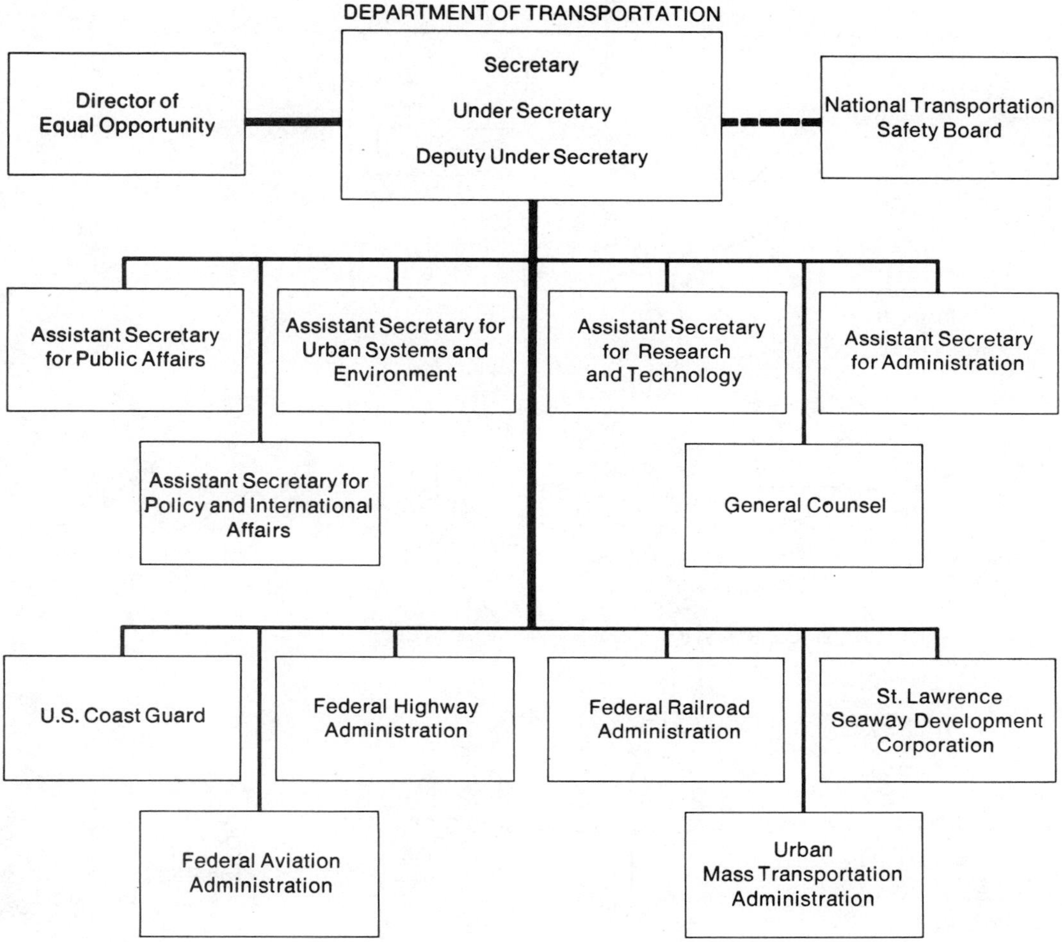

***Figure* 8.9**
Organization chart.[4]

---

[4] U.S. Department of Transportation (Washington, D.C.: U.S. Government Printing Office, December 1968), p. 24.

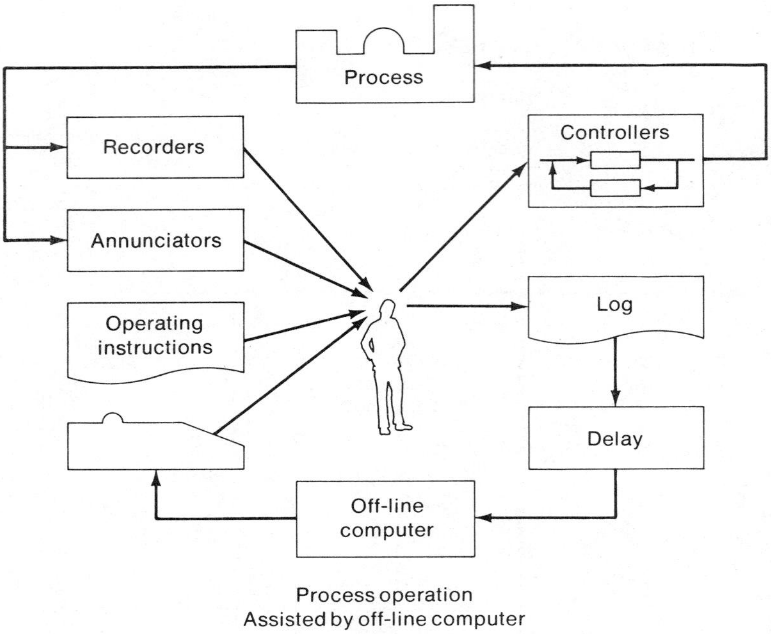

FLOW CHART ILLUSTRATION OF OFF-LINE COMPUTER

Figure 8.10
Flow chart (with geometric figures).

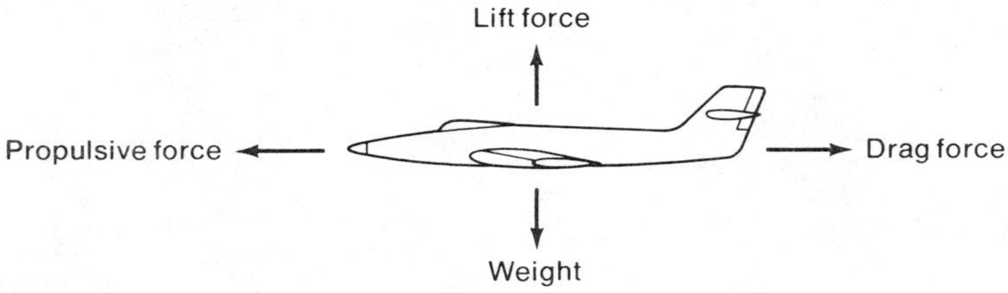

NEWTON'S THIRD LAW AS A PRINCIPLE OF AERODYNAMICS

Figure 8.11
Flow chart (with sketch).

Diagrams and Drawings

Like organizational charts and flow charts, diagrams and drawings are used to illustrate what the writer is explaining in his prose. In this text, the terms "diagram" and "drawing" will be used interchangeably. However, some of the important differences between them should be pointed out.

A drawing is a more fully fleshed-out picture of its subject. It closely resembles what the object it is depicting looks like. Figure 8.12 is an example of a drawing.

The word "diagram" usually refers to a skeletal sketch of the thing depicted. In other words, diagrams mainly show the structural framework of their subjects, but drawings more closely depict what they actually look like. Diagrams, therefore, are more suitable for showing how the subject functions or how it is used. Among the more common kinds are (1) line diagrams and (2) exploded diagrams.

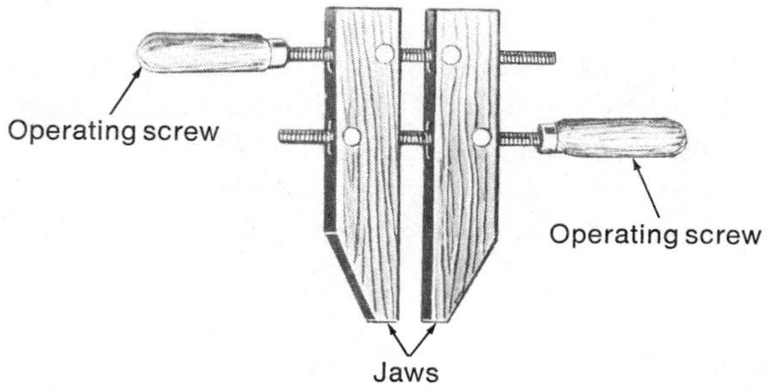

Figure 8.12
Drawing of a hand screw clamp.

Line diagrams can be prepared rapidly and easily. They are often used to enable the reader to see the structural relationships and dimensions. Line drawings also are used to enable the reader to visualize how an object works. Figure 8.13 is a structural line drawing of a kitchen sink, giving mainly the dimensions and design.

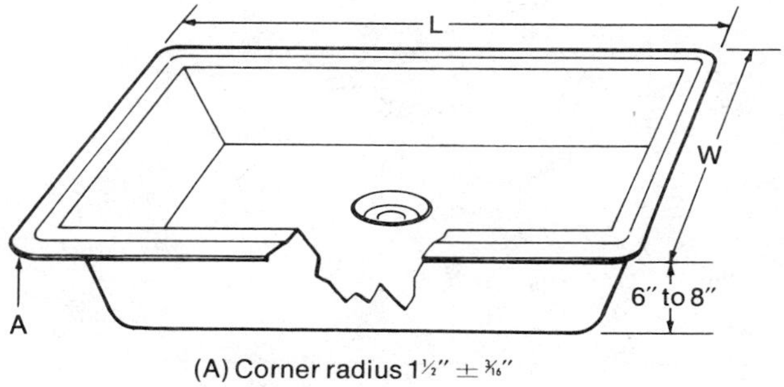

Figure 8.13
Line diagram of a flat-rim kitchen sink.

Figure 8.14 is a functional line drawing, mainly showing a process. It shows how salt water is first pumped through a filter, then through a desalting unit to make fresh water. It also lets the reader see how the parts are arranged.

Figure 8.14
Schematic diagram of the reverse-osmosis process.

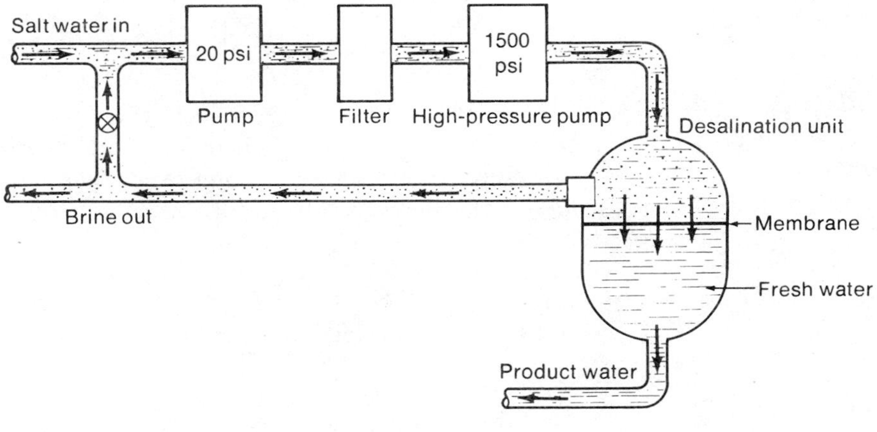

Exploded Diagrams

An exploded diagram or drawing is often used to show a reader how the parts of a device fit together, or how they should be assembled. Each part is drawn slightly separated from the others. Figure 8.15 illustrates how this type of diagram clearly explains the interrelationships between the parts as well as their assembly. Also note how each part is "called out" or clearly labeled.

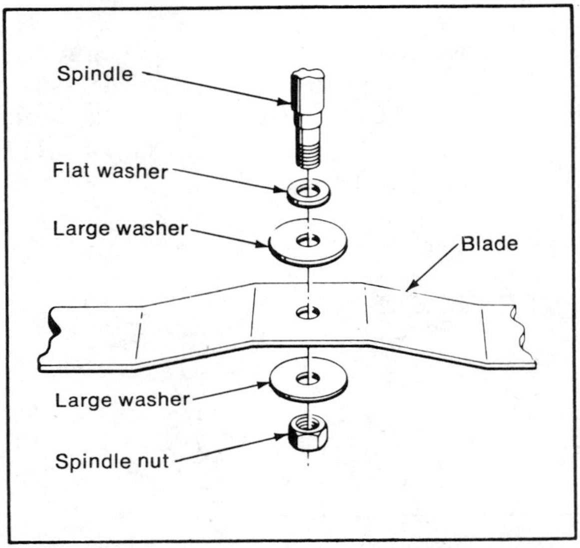

Figure 8.15
Exploded drawing of a lawnmower blade assembly.

Diagrams and drawings have distinct advantages over photographs. A diagram can be focused better by leaving out details that have little relevance to what the writer is explaining. Parts of the drawing or diagram can be colored, shaded, or crosshatched to make certain portions stand out. Inside portions of the subject can be shown without actually cutting it open. Diagrams and drawings are easy to alter, often merely by erasing, to enable them to express more clearly the prose explanations of the writer.

PHOTOGRAPHS

Even though photographs require more effort to prepare, they often have distinct advantages, making them useful in technical reports. They are especially useful when the writer needs to give the reader an exterior or overall view of the subject. They are used effectively, also, when the environment needs to be shown to give the photographed object a realistic appearance. Photographs can be retouched, cropped, or cut into desired shapes or portions.

Photographs

Two general types used in technical writing are on-location photographs and studio photographs. Studio photographs, of course, are taken under controlled conditions, usually in the professional photographer's studio, where ideal lighting, carefully determined range and angle, the most appropriate background details, and focus of the camera upon the features to be emphasized help to make the photograph an effective means of communication.

Although a technical writer is seldom required to use his own photographs in his reports, he is expected to be able to distinguish one type of photograph from another when reading the reports prepared by others. Therefore, it is worthwhile for him to be able to recognize some of the more common kinds of technical photographs: documentary, callout, and cutaway.

Documentary photographs show the subject in its normal environment. Therefore, they are inexpensive, and they are useful in giving the reader a realistic view of the subject. For example, a photograph taken at the scene of an automobile accident can be very helpful in substantiating what actually happened. Documentary photographs are also useful in revealing the form, dimensions, conditions, and overall appearance of something.

Callout photographs help a writer identify certain components of a subject. He inserts letters or numbers above or within the components that are to be identified and adds somewhere near the photographs a legend explaining what each letter or number represents. When there is room, however, it is better to place the identifications themselves within the photograph or around its perimeter, with an arrow directing the reader's eye to each part.

Cutaways are made by actually cutting a part of the device away and taking a picture. Of course, this may be very expensive. When it is too costly, a professional photographer or artist may be called upon to remove outer portions of the subject from the photograph and to insert a view of the inner parts. This gives the appearance that the outer portions of the subject were cut away, revealing its innards.

Application 8–14

Select any device with which you are familiar, perhaps the radiator of an automobile, and draw a flow chart.

Application 8–15

Search through magazines for examples of the diagrams we have discussed. Cut out one of each and paste it on a sheet of paper. Beneath one of them write a paragraph or two describing its shape, dimensions, ingredients or materials, etc.

Application 8–16

Draw a line diagram of a lighting fixture, a lamp, a chair, a flashlight, a mouse trap, or some other device with which you are familiar. Be sure to insert the names of the components and, if needed, their dimensions. You also may want to show the overall dimensions of the device. Also, below the diagram be sure to indicate the kind of diagram you intended along with the figure number and the title.

Application 8–17 Draw an exploded diagram of any of the items suggested in the preceding applications or some other item with which you may be more familiar. Remember, the purpose is to enable the reader to understand how the parts of the device look and how they fit together. Of course, the diagram also indicates the appearance of the overall device.

Application 8–18 Study the accompanying floor plan of a home showing a proposed addition, and write a realistic description based on the details. Be sure to maintain an objective viewpoint.

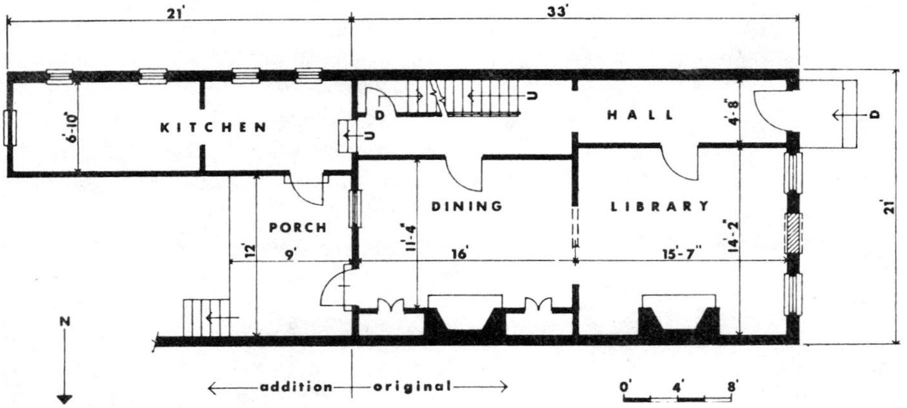

Application 8–19 Study the two flow charts below depicting the similarities and differences in the way electricity is generated in a nuclear plant and a fossil fuel plant, and write an explanation of it. Be sure to use a combination of comparison and contrast as well as any other kind of supporting form needed.

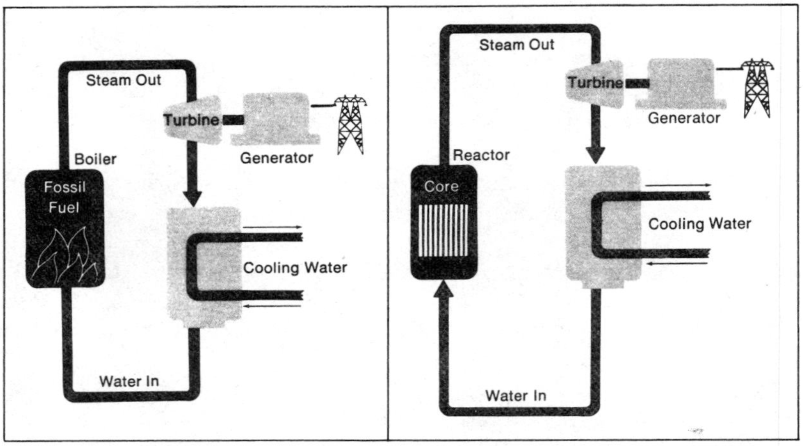

SOURCE: Energy Research and Development Administration, Division of Reactor Research and Development, "Advanced Nuclear Reactors" (Washington, D.C.: U.S. Government Printing Office, Sept. 1975), p. 2.

9

Reading better from writing better

how learning to write better helps reading . . . reading as communication . . . reading is translating . . . detecting main ideas and supporting ideas . . . seeing how facts, inferences, and opinions are used . . . determining the writer's dominant general purpose . . . identifying the intended writer/ reader relationship . . . seeing how basic forms of development are used: enumeration and analysis . . . observing how description, narration, and causation are developed by definition, illustration, classification, comparison, contrast, analogy . . . learning new words . . . witnessing the skillful use of words

People holding high professional or business positions often attribute their success to what others have told them in books and other forms of technical writing. These same people, however, may reach a point when they can't keep up with the large amount of new information added to their occupations. Even most students are constantly looking for ways to increase their reading speed, comprehension, and enjoyment. For these reasons many studies have been made recently to discover ways to increase reading skills.

This chapter will help you improve your reading by giving you both some theories about proficient reading and some applications in which you can practice them. You will get much satisfaction as you detect improvement in your reading speed and comprehension. More important, however, this improvement will continue long after you have left school.

Everything you have studied so far in this book about writing should help you to improve your reading. Being able to identify and distinguish the different types of ideas a writer puts into certain kinds of writing, you are better able to adapt your reading to different kinds of writing. When a writer is expressing entertaining opinions, you read one way, but when you know facts and reliable inferences are being expressed, you shift gears and read another way. For example, you read an entertaining short story one way and a technical report another.

If you can recognize the ways a writer develops and organizes ideas, you are able to continue reading much more smoothly because the relationships between ideas are clearer. A person able to draw blueprints should be better able to read them because the meanings of the markings are clearly understood. Therefore, what you have learned about the basic, general, and supporting forms of development will greatly contribute to your reading proficiency. Even learning to interpret nonverbal elements, such as graphs, charts, and drawings, should help you to read technical material with more understanding.

Reading is not just the taking in of words one right after the other. Like writing, reading is a process, a sequence of related steps taken to achieve a certain purpose, an objective. In Chapter 1 you were shown how to adapt what you say to the type of reader and to your intended purpose. To inform or persuade an expert, you express certain kinds of ideas and state them in a certain way. To do the same for the uninformed reader, you express different ideas in a different way.

To improve your technical reading, a good place to start is to learn to identify the writer's purpose and the type of reader his communication is aimed at. Once you know this, you can decide whether or not you can benefit enough from the communication to read it. If you decide that the material is not worthwhile, you just put it away. If you think it worth reading, you have to decide next how to read it. A good reader learns to adapt the kind of reading he or she does to the intended purpose of the writer and to his or her own purpose for reading it.

There are at least three different kinds of reading. In this book we call them (1) scan-reading, (2) skim-reading, and (3) study-reading.

SCAN-READING

Scanning for specific details is one kind of reading you should learn to do. Doing it skillfully will enable you to cover a large amount of material at a high rate of speed.

Scan-reading is done by focusing just on certain parts of a piece of writing. It is a kind of spot-checking in search of something. In doing it, the reader jumps from one spot to another. The next time you read a newspaper, notice carefully how you do it. You will discover that your eyes move rapidly up, down, and across, pausing only over a sentence or two here and there. People with busy office jobs read many routine letters that way. However, when they find something that requires a different kind of reading, they shift to skim-reading or study-reading.

Application 9–1 Scan the following for (1) a judgment as to which kind of bottle is better, (2) the conclusion reached, and (3) any recommendations.

BOTTLES FOR CARBONATED SOFT DRINKS [1]

The invention of automated machinery for soft drink bottles in 1907 introduced returnable glass bottles into the hands of the consumer. Returnable bottles were used by soft drink manufacturers until the complaints about handling the returnables began to be heard from retailers. In 1948, glass manufacturers introduced bottles of lighter weight, which still met the pressure and handling requirements of the trade—the nonreturnable bottle. The nonreturnable bottles—designed for a one-service trip—were lower in costs, decreased the work of the retailer, and were lighter in weight. According to the testimony of the Soft Drink Worker's Union at the hearings before the National Commission on Product Safety in 1969, the decreased thickness of glass in the nonreturnable bottle increased the likelihood of breakage.

The returnable continues to be manufactured but not in the same large quantities as nonreturnables. The returnable bottle is purported to have characteristics that permit multiple service trips as a carbonated soft drink bottle. Returnable bottles may withstand greater internal pressure than nonreturnable since their walls are made of thicker glass. During production they rest on a bearing ring and an extra thickness of additional strength. The bottling industry estimates that a returnable bottle makes an average of 16 round trips.

According to the trade association, the Glass Container Manufacturers Institute, over 10 billion carbonated soft drink bottles were produced for domestic use in 1973; 9 billion were nonreturnable bottles. In comparison to 1962, a little over 2 billion soft drink bottles were produced: 477 million nonreturnable and 1,574 million returnable bottles. The convenience of nonreturnable bottles and their lower manufacturing costs have contributed to their increased production.

In the production of carbonated soft drink bottles, the National Soft Drink Association estimates a business amounting to over $598 million in 1974. Carbonated soft drink bottles comprised 26 percent of all glass containers produced in 1973.

Problems surrounding conventional returnable and nonreturnable bottles have aroused the concern of product-safety experts. Incidents of exploding bottles, exploding bottles resulting from impact, breakage resulting from impact, propulsion of bottle caps, and accidental contact with broken glass are shown in this analysis to be associated with carbonated beverage bottles. It has not been determined which type of carbonated soft drink bottle is more hazardous—returnable or nonreturnable. The returnable bottle may be subjected to more round trips and may receive abuse during repeated trips between bottler, store, and consumer.

There also is concern, however, about using large bottles (i.e., the 32 fluid ounce, the 48 fluid ounce, and the 64 fluid ounce bottles) for

---

[1] U.S. Consumer Product Safety Commission, Bureau of Epidemiology, *Hazard Analysis: Bottles for Carbonated Soft Drinks,* March 1975, p. 1.

carbonated beverages. There are greater amounts of potential energy in the form of compressed gas in these larger bottles. Therefore, in the event of breakage or failure, this extra energy can propel fragments with more force. The larger amount of glass used also adds to the material available to be transformed into sharp missiles.

Regardless of the type of bottle, the question of whether they are being adequately inspected in the processing line is being considered. After the initial testing by the glass manufacturer for internal pressure strength, abrasion resistance, visual defects, thermal shock resistance, as well as other factors, the bottle may not be tested again by the bottler, carrier, or the distributor.

SKIM-READING

Skimming, like scanning, does not require you to see and take in every word on the page. It is slower, however, than scanning. Skim-reading is a "once-over-lightly" kind of reading. To get a general idea what a report is about and the kinds of supporting ideas it contains, you may choose to skim it.

In scan-reading you usually search for something more specific, but in skim-reading you sift through the whole report lightly, looking for main ideas, skimming off main details, or just sampling the quality of the writing.

Application 9–2

Skim through the following paragraphs and identify the writer's purpose, to inform or to persuade. Next, write a sentence or two summarizing the main ideas in each paragraph.

1. What are the most common sources of injury during grinding operations? Hazards leading to eye injury caused by grit generated by the grinding process are the most common and the most serious. Abrasions caused by bodily contact with the wheel are quite painful and can be serious. Cuts and bruises caused by segments of an exploded wheel, or a tool "kicked" away from the wheel, are other sources of injury. Cuts and abrasions can become infected if not protected from grit and dust from grinding.
2. The utilization of bark for mulch and soil conditioner represents a practical environmental advance. As much as 15 percent of a tree's volume is bark. Until very recently, the processing of this bark into useful products was considered both costly and unnecessary. Trees taken to the sawmill were first stripped of their bark. While their logs went on to become lumber products, their bark was cast aside. Because it accumulates quickly (about 1,200 pounds of green bark are by-products of each 1,000 board feet of lumber manufactured), bark presents problems and is a potential fire hazard. Sawmill operators chose their only practical alterna-

tive, which was to dispose of the useless and troublesome bark quickly by burning it in tepee burners. Naturally this contributed to air pollution.

3. In general, a fire whirl is a stronger and more intense dust devil—in characteristics and mode of development the two are much alike. Both depend on a supply of heated, buoyant air. On hot days when the skies are clear and winds light, dry land surfaces can become strongly heated by the sun—surface temperatures of 165° F are not uncommon. Under these conditions, a layer of superheated air is formed near the surface. This hot, buoyant surface air tends to rise in columns, which draw in more of the hot surface air as they move along with the wind. Some of the columns develop a strong rotational motion—these pick up dust and debris and become visible as a dust devil.
4. People lose some of their freedom by demanding and installing more mechanization in their homes. Doing this makes them more and more dependent upon factory-produced items: refrigerators, electronic ovens, washers and dryers, heating and cooling systems, and a multitude of personal-care devices. These are provided and regulated by state and city governing agencies. Besides producers, consumers, and those who provide the utilities, another force to be reckoned with is the regulator of housing. Building codes, zoning orders, and subdivision controls are needed. Beyond these are newer agencies concerned with the quality of the environment—pollution monitors, licensing agencies, and environmental protection programs.
5. The art of subsurface excavation has been known to mankind for many centuries. Archaeologists reported that subsurface mining began during the Stone Age, some 15,000 years ago. Evidence of a number of these mines has been found in Europe. After the surface supply of flint nodules, which were used for weapons and tools, was exhausted, the flint outcrop was followed into the ground, first in pits, then later through shafts from which headings were driven into the deposits. The flint miners used picks made from deer antlers and flint axes to dislodge the chalk-embedded nodules and then loaded the material with shovels made from the shoulder blades of horses. The splitting action of the wedge was mastered at this early date, as the chalk was split by hammering wedges made of horses' bones.
6. Foods made from all-vegetable components that resemble forms of beef or poultry or other animal products have been sold in the United States for many years. Meat "extenders," such as soy granules, are combined with animal protein and "spun" into long, thin fibers, then pressed into shapes resembling meat, such as "ham-like" cubes, "pepperoni" slices, or "bacon-like" pieces. By a less expensive method they can be "extruded" under pressure into chips, flakes, chunks, and a variety of other shapes.

Application 9–3

Skim through the following article and identify the writer's purpose. Next, write a brief summary of the main ideas in the whole article. Also, list at least two facts, inferences, and opinions.

SPACE SHUTTLE: Emphasis for the 1970's [2]

After a decade highlighted by driving effort and dramatic achievement, America's space program is shifting emphasis. Now the goal is practical benefits for people on Earth.

The first 15 years of the Space Age have witnessed a vast outpouring of new knowledge, and development of new technology, skills and whole areas of science and engineering. These point to answers for problems that could not be solved without space flight.

Even at this early date, the practical benefits from the space program far exceed the costs. Evacuations saved thousands of lives in 1969 when weather satellites forewarned that Hurricane Camille would slam into the Gulf Coast. Communications satellites now carry half the world's international telephone, telegraph and television traffic at substantially lower prices than those of a decade ago. Thousands of new employment opportunities have been created in areas such as the $8 billion-a-year computer industry.

Soon, satellites will sense air and water pollution, send warnings of crop disease, scan the oceans for the best fishing areas and search the Earth for geologic formations associated with untapped oil and mineral reserves. The accuracy of weather forecasts will extend from the present one or two days to one or two weeks by the late 1970s. Space navigation aids will enhance flight safety in the airlanes between America and Europe. Further technological progress will lead to more jobs for Americans and increased sales of U.S. products overseas.

But, barriers stand in the way. Before these benefits can be achieved there must be breakthroughs in cost, time and simplicity. Space flight is still expensive. Manufacturing, testing and launching of satellites is extremely costly. Five or six years may elapse before the idea in a scientist's mind becomes the reality of a flight experiment; more time is needed to develop a satellite into a practical tool. Scientists and engineers must use remote controls. They cannot retrieve a satellite for trouble shooting if it does not operate as expected. The expensive booster for each flight to space can only be used once.

These barriers will be broken by the Space Shuttle, a revolutionary new vehicle that will combine the advantages of airplanes and spacecraft, and will fly repeatedly to space and back to Earth. It will not be expended as present space vehicles are after a single flight. Many millions of dollars will be saved by using satellite equipment over and over again, and by using low-cost standard components that can be replaced when they wear out. The years of preparation for space flight will be dramatically shortened.

Technicians and specialists will accompany satellites into orbit, make adjustments as necessary, and bring them back to Earth for modernization and maintenance. Thus, the introduction of practical benefits will become economical, speedy and simple.

The Space Shuttle will take off vertically with a pilot and a co-pilot at the helm and two other crew members. In early operations, the Shuttle port will be at Kennedy Space Center, Florida, for east-west orbits. Later a port will be added at Vandenberg Air Force Base, California, for north-south orbits. Two solid-propellant booster rockets will supply most of the takeoff power. About 40 kilometers (25 miles) high, the boosters will separate and descend by parachute to the ocean surface. There they will be recovered and returned to the launch site for re-use.

The main section of the Shuttle, called the Orbiter, will continue flying on the power of its liquid-propellant engines, supplied by a large external tank. After these two sections reach orbit, the tank will separate and a small rocket will cause it to re-enter and land in a remote ocean area. The Orbiter will be able to carry out space missions lasting at least seven days. Special materials covering its entire surface will protect the interior from the searing heat of re-entry. The Orbiter will fly horizontally like an airplane during the latter phase of descent and it will land on a runway near the launch site. As ground crews gain experience in readying it for subsequent flights,

---

[2] National Aeronautics and Space Administration, Office of Public Affairs, "Space Shuttle: Emphasis for the 1970's" (Washington, D.C.: U.S. Government Printing Office, June 1972).

the turnaround time will be reduced to two weeks.

When operational, the Space Shuttle will replace all but the smallest U.S. space launch vehicles. It will launch and return weather satellites, communications satellites, pollution control satellites, Earth resource satellites, navigation satellites, scientific satellites and space probes. It will provide launch services for the Department of Defense and other agencies of the U.S. Government, foreign countries, private industry and research organizations. It will operate as a common carrier, serving essentially anyone who can buy a ticket or pay the freight cost.

Passengers need not necessarily meet the present stringent physical standards for space flight. They may be scientists, engineers, technicians, journalists, television crews or others whose business takes them into space. As experience increases the assurance of safety, men and women of many organizations and many countries will be among the passengers.

As costs decrease, preparation times shorten and operations are simplified, new uses of space flight will develop. Among those now envisioned are the manufacture of high-cost, high-purity products like vaccines, exotic metal alloys and special castings. Industrial researchers expect the weightlessness of space will lead to economic advantages that will warrant the cost of space activity.

Scientists are also considering how to collect the Sun's energy in space and convert it to electrical power for transmission to Earth without pollution.

But scientific leaders believe the most significant benefits to people on Earth will come from inventions not yet conceived, which will be stimulated when the Shuttle makes space flight simple, less time consuming and less expensive.

Application 9–4

The words below are used by Mortimer Adler in "How to Mark a Book," which follows this application. Each word is used in the paragraph identified by the number next to it. First, write the word, then read the paragraph in which it is placed and write opposite it what you think it means from the way it is used. Next, check for the appropriate meaning in a dictionary and write it opposite your own interpretation.

mutilation (2)	maestro (8)	integrated (23)
prelude (4)	invariably (11)	protrude (24)
restrain (6)	integral (13)	laboriously (25)
unblemished (7)	sparingly (18)	

Now scan-read the article to locate the paragraphs in which you find:

1. An analogy between beefsteak and owning a book.
2. A classification of kinds of people who have books.
3. The title of a famous novel.
4. The name of a well-known college president.
5. The name of a famous English writer.

Application 9–5

Skim-read "How to Mark a Book" to decide whether Adler's purpose is to inform or to persuade. If you think it is to inform, write "to inform," and under it list the main bits of new information you learned from reading the whole article. If you think the author's main purpose is to persuade you, list the main reasons he gives to change your beliefs about how a book should be read. Skim out only two or three of the most important main reasons.

HOW TO MARK A BOOK [3]

[1] You know you have to read "between the lines" to get the most out of anything. I want to persuade you to do something equally important in the course of your reading. I want to persuade you to "write between the lines." Unless you do, you are not likely to do the most efficient kind of reading.

[2] I contend, quite bluntly, that marking up a book is not an act of mutilation but of love.

[3] You shouldn't mark up a book which isn't yours. Librarians (or your friends) who lend you books expect you to keep them clean, and you should. If you decide that I am right about the usefulness of marking books, you will have to buy them. Most of the world's great books are available today, in reprint editions, at less than a dollar.

[4] There are two ways in which one can own a book. The first is the property right you establish by paying for it, just as you pay for clothes and furniture. But this act of purchase is only the prelude to possession. Full ownership comes only when you have made it a part of yourself, and the best way to make yourself a part of it is by writing in it. An illustration may make the point clear. You buy a beefsteak and transfer it from the butcher's icebox to your own. But you do not own the beefsteak in the most important sense until you consume it and get it into your bloodstream. I am arguing that books, too, must be absorbed in your bloodstream to do you any good.

[5] Confusion about what it means to *own* a book leads people to a false reverence for paper, binding, and type—a respect for the physical thing—the craft of the printer rather than the genius of the author. They forget that it is possible for a man to acquire the idea, to possess the beauty, which a great book contains, without staking his claim by pasting his bookplate inside the cover. Having a fine library doesn't prove that its owner has a mind enriched by books; it proves nothing more than that he, his father, or his wife, was rich enough to buy them.

[6] There are three kinds of book owners. The first has all the standard sets and best-sellers—unread, untouched. (This deluded individual owns woodpulp and ink, not books.) The second has a great many books—a few of them read through, most of them dipped into, but all of them as clean and shiny as the day they were bought. (This person would probably like to make books his own, but is restrained by a false respect for their physical appearance.) The third has a few books or many—every one of them dog-eared and dilapidated, shaken and loosened by continual use, marked and scribbled in from front to back. (This man owns books.)

[7] Is it false respect, you may ask, to preserve intact and unblemished a beautifully printed book, and elegantly bound edition? Of course not. I'd no more scribble all over the first edition of *Paradise Lost* than I'd give my baby a set of crayons and an original Rembrandt! I wouldn't mark up a painting or a statue. Its soul, so to speak, is inseparable from its body. And the beauty of a rare edition or of a richly manufactured volume is like that of a painting or a statue.

[8] But the soul of a book *can* be separated from its body. A book is more like the score of a piece of music than it is like a painting. No great musician confuses a symphony with the printed sheets of music. Arturo Toscanini reveres Brahms, but Toscanini's score of the C-minor Symphony is so thoroughly marked up that no one but the maestro himself can read it. The reason why a great conductor makes notations on his musical scores—marks them up again and again each time he returns to study them—is the reason why you should mark up your books. If your respect for magnificent binding or typography gets in the way, buy yourself a cheap edition and pay your respects to the author.

[9] Why is marking up a book indispensable to reading it? First, it keeps you awake. (And I don't mean merely conscious; I mean wide awake.) In the second place, reading, if it is active, is thinking, and thinking tends to express itself in words, spoken or written. The marked book is usually the thought-through book. Finally, writing helps you remember the thoughts you had, or the thoughts the

[3] Mortimer Adler, "How to Mark a Book," *Saturday Review of Literature,* July 6, 1940.

author expressed. Let me develop these three points.

[10] If reading is to accomplish anything more than passing time, it must be active. You can't let your eyes glide across the lines of a book and come up with an understanding of what you have read. Now an ordinary piece of light fiction, like say, *Gone with the Wind*, doesn't require the most active kind of reading. The books you read for pleasure can be read in a state of relaxation, and nothing is lost. But a great book, rich in ideas and beauty, a book that raises and tries to answer great fundamental questions, demands the most active reading of which you are capable. You don't absorb the ideas of John Dewey the way you absorb the crooning of Mr. Vallee. You have to reach for them. That you cannot do while you're asleep.

[11] If, when you've finished reading a book, the pages are filled with your notes, you know that you read actively. The most famous *active* reader of great books I know is President Hutchins, of the University of Chicago. He also has the hardest schedule of business activities of any man I know. He invariably reads with a pencil, and sometimes, when he picks up a book and pencil in the evening, he finds himself, instead of making intelligent notes, drawing what he calls "caviar factories" on the margins. When that happens, he puts the book down. He knows he's too tired to read, and he's just wasting time.

[12] But, you may ask, why is writing necessary? Well, the physical act of writing, with your own hand, brings words and sentences more sharply before your mind and preserves them better in your memory. To set down your reaction to important words and sentences you have read, and the questions they have raised in your mind, is to preserve those reactions and sharpen those questions.

[13] Even if you wrote on a scratch pad, and threw the paper away when you had finished writing, your grasp of the book would be surer. But you don't have to throw the paper away. The margins (top and bottom, as well as side), the end-papers, the very space between the lines, are all available. They aren't sacred. And, best of all, your marks and notes become an integral part of the book and stay there forever. You can pick up the book the following week or year, and there are all your points of agreement, disagreement, doubt, and inquiry. It's like resuming an interrupted conversation with the advantage of being able to pick up where you left off.

[14] And that is exactly what reading a book should be: a conversation between you and the author. Presumably he knows more about the subject than you do; naturally, you'll have the proper humility as you approach him. But don't let anybody tell you that a reader is supposed to be solely on the receiving end. Understanding is a two-way operation; learning doesn't consist in being an empty receptacle. The learner has to question himself and question the teacher. He even has to argue with the teacher, once he understands what the teacher is saying. And marking a book is literally an expression of your differences, or agreements of opinion, with the author.

[15] There are all kinds of devices for marking a book intelligently and fruitfully. Here's the way I do it:

[16] *1. Underlining:* Of major points, of important or forceful statements.

[17] *2. Vertical lines at the margin:* To emphasize a statement already underlined.

[18] *3. Star, asterisk, or other doo-dad at the margin:* To be used sparingly, to emphasize the ten or twenty most important statements in the book. (You may want to fold the bottom corner of each page on which you use such marks. It won't hurt the sturdy paper on which most modern books are printed, and you will be able to take the book off the shelf at any time and, by opening it at the folded-corner page, refresh your recollection of the book.)

[19] *4. Numbers in the margin:* To indicate the sequence of points the author makes in developing a single argument.

[20] *5. Numbers of other pages in the margin:* To indicate where else in the book the author made points relevant to the point marked; to tie up the ideas in a book, which, though they may be separated by many pages, belong together.

[21] *6. Circling of key words or phrases.*

[22] *7. Writing in the margin, or at the top or bottom of the page, for the sake of:* Recording questions (and perhaps answers) which a passage raised in your mind; reducing a complicated discussion to a simple statement; recording the sequence of major points right through the book. I use the end-papers at the back of the book to make a personal index

of the author's points in the order of their appearance.

[23] The front end-papers are, to me, the most important. Some people reserve them for a fancy bookplate. I reserve them for fancy thinking. After I have finished reading the book and making my personal index on the back end-papers, I turn to the front and try to outline the book, not page by page, or point by point (I've already done that at the back), but as an integrated structure, with a basic unity and an order of parts. This outline is, to me, the measure of my understanding of the work.

[24] If you're a die-hard anti-book-marker, you may object that the margins, the space between the lines, and the end-papers don't give you room enough. All right. How about using a scratch pad slightly smaller than the page-size of the book—so that the edges of the sheets won't protrude? Make your index, outlines, and even your notes on the pad, and then insert these sheets permanently inside the front and back covers of the book.

[25] Or, you may say that this business of marking books is going to slow up your reading. It probably will. That's one of the reasons for doing it. Most of us have been taken in by the notion that speed of reading is a measure of our intelligence. There is no such thing as the right speed for intelligent reading. Some things should be read quickly and effortlessly, and some should be read slowly and even laboriously. The sign of intelligence in reading is the ability to read different things differently according to their worth. In the case of good books, the point is not to see how many of them you can get through, but rather how many can get through you—how many you can make your own. A few friends are better than a thousand acquaintances. If this be your aim, as it should be, you will not be impatient if it takes more time and effort to read a great book than it does a newspaper.

[26] You may have one final objection to marking books. You can't lend them to your friends because nobody else can read them without being distracted by your notes. Furthermore, you won't want to lend them because a marked copy is a kind of intellectual diary, and lending it is almost like giving your mind away.

[27] If your friend wishes to read your *Plutarch's Lives*, "Shakespeare," or *The Federalist Papers*, tell him gently but firmly, to buy a copy. You will lend him your car or your coat—but your books are as much a part of you as your head or your heart.

STUDY-READING

Studying requires a slow, thorough kind of reading. When you study-read something, you try to understand not only the individual main ideas but also the relationship between them. While doing this kind of reading, you may even make your own interpretations, draw your own inferences, from the ideas stated. Your interpretations may or may not agree with those of the writer.

Just as in scanning and skimming, you adjust the kind of reading you do to your purpose. When your purpose is to study-read, you focus your eyes and especially your mind sharply as you closely examine the main and the minor ideas. Knowing your purpose, your mind almost automatically coordinates all the complex skills needed to study-read efficiently.

Application 9–6 Go back and study-read "How to Mark a Book" and write the main idea for the whole article. Listed below are the numbers of some of the article's main paragraphs with the main idea for each opposite it. Study-read each corresponding paragraph and, under its main idea, list the ideas Adler uses to support it.

Paragraph	*Adler's main idea*
4	There are two ways in which one can own a book.
6	There are three kinds of book owners.
8	But the soul of a book *can* be separated from its body.
9	Why is marking up a book indispensable to reading it?
12	But, you may ask, why is writing necessary?
14	And that is exactly what reading a book should be: a conversation between you and the author.
25	Or, you may say that this business of marking books is going to slow up your reading.

Application 9–7 Write a one- or two-paragraph summary of "How to Mark a Book," giving the main supporting ideas and conclusions. Be sure to include the specific steps Adler gives for marking a book.

Application 9–8 1. The following words are from the article below, "Cities in the Sky." Without reading the article or consulting a dictionary, write a synonym and antonym for each. Then while reading the article find each word and determine from the context whether your synonyms and antonyms are correct. If you need to do so, check a dictionary to decide about the correctness of your words.

expanse	habitat
feasible	sophisticated
hypothetical	unique
implementing	linear
prototype	cumbersome
biosphere	centrifugal
counter-rotate	conventional
opaque	

2. Scan-read the article and find the exact definition of a "liberation center."

CITIES IN THE SKY [4]

Imagine a huge cylinder four miles in diameter and about 16 miles long orbiting in outer space. Inside, you step from your cozy little cottage, stroll down a flower-lined garden path, wander through bird-filled wildlands, picnic in a forest park, swim in a lake, fish in a river. Later, you hop on your bicycle or in your tiny electric car and run down the street to the shopping center for some groceries.

The cylinder will be rotating to produce artificial gravity, but you won't notice the motion. Everything will be just as it is on Earth—with a few pleasant exceptions. There will be no bustling crowds, no traffic jams, no blaring car horns, no ear-shattering sounds of heavy industry, no eye-smarting smog. The air will be clean and fresh and the climate perfectly controlled—except it may rain now and then for there are real clouds up in the "sky."

If you like mountain climbing, you can scale peaks up to 8000 feet high and—oddly enough—your backpack will get lighter and the going easier the higher you go because the artificial gravity decreases as you approach the center of the cylinder. And if you want to soar like a bird, just launch your pedal-powered glider from a hillside and off you sail—the lessened gravity at the "higher" altitudes will make man-powered flight easily possible.

There will be other strange effects, too. As you look across the four-mile expanse of the cylinder's interior, your neighbors' homes will appear upside down on the "roof." Water will run uphill. If it decides to rain, it will rain

---

[4] Richard Dempewolff, "Cities in the Sky," *Popular Mechanics*, May 1975, pp. 94–97,205. Reprinted by permission of *Popular Mechanics*, Copyright © 1975 The Hearst Corporation.

in all directions—up and down at the same time. Up on those mountains there will be special trick "low-gravity" swimming pools where you can dive off a board and take several seconds before hitting the water. When you aren't swimming, you'll be able to bask on beaches and get a real suntan from real sunlight streaming in through long, portlike slots in the sides.

If all this sounds like science fiction, it isn't. You're living in a man-made mini-world—a bold new plan to put entire human colonies into space. Not just orbiting space stations filled with antiseptic lab gear, but actual full-scale reproductions of Earthlike communities complete with houses, schools, stores, hills and valleys, trees and streams, roads and wildlife.

The project is not only feasible; it is under serious study right now. This month some 100 prominent physicists, engineers, astronomers, space-flight experts and other scientists will meet at Princeton University to confer on realistic—not hypothetical—ways and means of implementing the plan. Organizer of the gathering and originator of the space-colony proposal is Dr. Gerald K. O'Neill, professor of physics at Princeton. O'Neill is a young, personable, dynamic, entirely hardheaded physicist who knows what he's about. "We can colonize space now," he says, "using 1970s technology. New techniques and approaches are needed, but knowledge and capability exist." He has data to prove it.

The space colony concept began as an "amusing" exercise for a group of ambitious physics students. The object, says the professor, was to make otherwise dull calculations more interesting by applying them to a glamorous, if theoretical, project. "As sometimes happens in the hard sciences," he observes, "what began as a joke had to be taken seriously when the numbers began to come out right."

The numbers indicated that completely self-contained, self-supporting, highly livable man-made mini-worlds could, indeed, be constructed in space—using *only* materials and technologies that already exist!

The initial prototype colony, O'Neill suggests, would be a small one supporting about 10,000 people. From this would eventually come full-size space communities with populations of 400,000 circling the Earth. If a decision is reached to begin construction soon, O'Neill estimates the first colony could be completed by the late 1980s—though you probably won't be able to move right in. It will take several years, using electrical power from solar energy, to get the gigantic cylinders spinning at proper speed. In any case, space communities *could* be ready by about the end of the century—in time for your children to be among the first space settlers in history.

Assuming free use of the present body of technical knowledge, projected cost of the initial prototype colony is comparable to that of the Apollo project—about 33 billion 1972 dollars. Within a few years, however, products and power produced in space would pay for the cost of construction, thus making such colonies financially self-liquidating.

For our own Earth, the environmental and ecological effects of space colonization will be staggering. It will take the pressure off our current population explosion. It will lessen the drain on our rapidly dwindling natural resources. It will end famine and disease caused by overcrowding. It will eliminate industrial pollution of our air and waters because manufacturing and power-producing plants will be moved from Earth to special "clean" facilities in the sky as part of the space-colonization program.

If O'Neill's proposals are followed, here are some likely prospects for the future: By the late 1990s or early 2000s, there would be no need for the construction of additional power-generation plants on Earth. Solar power stations in orbit would beam energy to Earth by low-density microwave, avoiding any need for nuclear power. By 2015, a significant fraction of all humanity would be living in space colonies. The population on Earth might even begin to decline. A thriving economic interchange between Earth and space would be established. By 2050, many industries now polluting the Earth's biosphere would have been moved to space factories, using pollution-free technology. Waste would be recycled, employing the unlimited low-cost energy available in the colonies.

With its population stabilized and pollution eliminated, Earth could become the "Garden of Eden" of the solar system.

Meanwhile, life in space will be much like it is here, with regular day and night cycles, natural sunlight, normal gravity and an Earthlike environment. All this will take place inside O'Neill's four-mile-diameter cylinders whirling around the Earth like satellite moons. The huge tubes will be pressurized to produce a "shirt-sleeve" atmosphere, and revolve to set up centrifugal forces simulating normal Earth gravity.

Actually, the cylinders will be arranged in pairs, tethered together by cables about 50 miles apart to insure precise orientation in space. The two tubes will counter-rotate, each canceling out any gyroscopic effect the other might produce, so the pair always remains pointed endwise toward the sun. The cylinders will each turn once every two minutes, powered by solar energy.

Outer walls of each cylinder will be divided into six longitudinal sections of alternating opaque and transparent material. The three opaque strips are "land" areas, built up with real rock and soil. Virtually any type of terrain can be simulated, from rolling Berkshire hills to sweeping western plains, a lush tropical paradise or lonely arid desert. Each land area is two miles wide by 16 miles long—big enough O'Neill points out, to duplicate the south coast of Bermuda or the Carmel Bay area of California. In the protoype plan, an 8000-foot mountain profile copied from the Grand Tetons in Wyoming is proposed as a surface for one cylinder end cap.

The alternating transparent panels, identical in size to the land strips, will admit sunlight. Three mammoth rectangular mirrors, hinged to open out above the transparent strips, will reflect the sun into interior land areas and make the sun visible to space dwellers just as it appears on Earth. By adjusting angles of the mirrors, it's possible to make the sun traverse the sky in an apparent dawn-to-dusk cycle and also to alter seasonal climates within the space colonies. You could even plan, suggests O'Neill, "a phase difference of seasons between the twin cylinders and have midwinter in one while it's midsummer in the other." Thus space colonists, visiting each other in small shuttle craft, could enjoy brief climate changes just as winter-bound Earthlings head south for a sunny retreat.

Almost nothing is impossible in the never-never storybook land of outer space. There *is* one problem, though—it's with stars. If you turn the mirrors away from the sun to reflect in the stars, the stars would appear to rotate every two minutes and "you'd get dizzy," O'Neill observes. But space dwellers will still get to see magnificent views of the stars from zero-gravity observatories and hotels just outside the habitat.

Crops will be grown outside the main cylinder in a "halo" of smaller cylinders or pods surrounding one end. The advantage of keeping agricultural areas outside, in addition to creating more living space inside, is that each one can be individually tailored to suit the specific things you want to grow. According to Dr. Eric Hannah, O'Neill's right-hand man at Princeton, "In each of those outside cylinders you can change the climate and seasons to accommodate the crop. Each is a separate environment geared to citrus fruit, poultry, winter wheat, livestock—whatever you want."

Other external pods would be used to house industries, workshops, automated factories, scientific research labs, observatories and other special facilities. Thus all industrial, manufacturing and experimental work would be kept safely away from "civilization" in the main colony. As Hannah explains, "The external pods would be lashed in a space frame of light cable, probably, so they wouldn't run into each other. But basically they'd be independent units. That way, if there's a plant or animal disease, you can isolate the unit and sterilize it." Products produced in these external labs and factories could be sold to Earth, including some highly sophisticated things that are actually easier to make in space such as special crystals, supermagnets and other high-vacuum items. Power from satellite stations could also be sold to Earth.

The space colonies will circle the Earth along the same orbital path as the moon. The reason is curious. It has long been known there are five orbiting points in space where the gravitational forces of Earth and moon balance out. Any object stationed within one of these vast, roughly spherical gravity-less "pockets" would tend to be trapped there, orbiting Earth forever without need for propulsion.

These five gravity-free points, known as "libration centers," are designated "L1" through "L5". Areas L4 and L5 follow the same orbit as the moon—L4 ahead of it and L5 following it. Of the five centers, only L4 and L5 are stable so these have been chosen as locations for the space colonies. The twin-cylinder space habitats would be strung out along these orbital paths at 120-mile intervals—far enough apart to be safely separated, yet close enough for easy access to each other. For intercolony travel, there would be public space shuttle "buses" making regular trips between habitats. To visit your nearer neighbors in your twin colony, you'd hop into an engineless, pilotless craft that can "unlock" from the outer surface of the revolving habitat at a preprogrammed instant and be flipped into space by centrifugal force at about 400 mph. In nine minutes, you'd "lock" onto the spinning adjacent cylinder at zero relative velocity. Docking towers and spaceports are at the "nose" of each cylinder to handle such craft, also space flights from Earth.

Most fantastic of all, perhaps, is the fact that these colossal colonies will not only be built *in* space, but will largely use material and energy sources *found* in space. Except for certain materials and special construction machinery needed initially to get the program started, only about two percent of all supplies will come from Earth, predicts O'Neill. Ore from the moon and asteroids, for instance, will be used extensively in fabricating parts of the habitats—at a fraction of the cost of moving the same materials from Earth. "To bring a kilogram of material from the moon to L5," says O'Neill, "takes less than five percent of the energy needed to move it from Earth."

To transport moon ore to construction sites in space, a unique device has been proposed. It takes advantage of the moon's vacuum and low escape velocity—less than a quarter of what would be required to boost comparable loads from Earth. This machine, the "Transport Linear Accelerator" (TLA), is a "mass driver," as O'Neill calls it. It is similar to a linear accelerator of nuclear particles. Instead of particles, however, it uses solar electric power to accelerate buckets of compacted moon ore along a six-mile course until they reach escape velocity, then slings them out into space toward a colony under construction. At the construction site, pickup vehicles would scoop up the free-flying bundles of ore and bring them in to the colony.

Another advantage of the system, notes O'Neill with a gleam, is that it also can serve as a reaction motor—it not only can hurl loads in one direction, but using the loads themselves as a reaction mass can also push something in the opposite direction. "Using any kind of moon slag or asteroidal debris as a reaction mass," says O'Neill, "it can be used to propel payloads in the 100,000-ton range or higher."

Imagine what this could mean. In the asteroid belt, for example, there are great clods of pure nickel-iron floating around that could easily be processed into steel if they could be moved to an L5 construction site. Many meteoroids are thought to be rich in hydrocarbons and water—both essential to a space colony. Hitch a TLA to one of these chunks and, by spewing out bits of it in one direction, you could shove it—even an entire asteroid—in the opposite direction toward your colony site.

To make construction in space possible without the need for cumbersome space suits, scientists have devised giant aluminum spheres that can be pressurized and heated to permit work inside in a shirt-sleeve environment. Parts for a colony would be fabricated one at a time in modules inside the spheres. As each module is completed, the sphere around it would split apart in two halves to permit its removal, then the halves would close again for work to begin on the next module.

Basic construction of a typical colony cylinder would consist of a web-like framework of cables and ribs—the cables running lengthwise to hold the four-mile-diameter end caps in place against the internal atmospheric pressure and ribs or cables running circularly around the circumference to contain the pressure and centrifugal force against the sides. Over this skeletal framework would be fastened steel, aluminum or titanium plates to form the opaque walls for the land areas. The transparent "window" strips in between would be glass panels reinforced by woven steel mesh.

Each colony would have a solar power station of paraboloidal mirrors encircling the after end of the cylinder, focusing the heat on boiler tubes. Conventional turbine generators would provide the entire community with a potential of 10 times the power per person available on Earth. Oxygen from plentiful lunar oxides would be combined with liquid hydrogen to produce water.

Inside each colony, population density would be low since all industrial and agricultural activities are kept outside. There would be about 100 square miles of forest and parkland to provide pleasant surroundings. Since the entire community would be heated as a unit by solar energy, there would be no need for individual heating systems. No fuel-burning cars would be allowed—only bicycles and small pollution-free electric cars.

As for sports, you'd have everything you can do on Earth—plus a few you can't do here. In wintertime, snow would fall for skiing in the mountains—mountains, incidentally, created from moon rock and soil. All sorts of trick sports involving low gravity could be performed at the higher levels where gravity decreases or in special pods outside the main cylinder. Skin diving in space would be a unique experience. This could be done, notes O'Neill, in one of the external pods filled with water and fish. Slow rotation would produce low gravity, thus eliminating pressure-equalization problems. Noisy and pollution-producing sports—like car racing—would also be possible, but would be confined to pods in the outer ring.

What about hazards of radiation and meteoroid damage? No problem, say the scientists. The metal skin, soil and the depth of atmosphere will take care of cosmic rays. As for a meteoroid striking a colony, the chances are slim—once in a million years for a one-ton chunk to collide is the estimate of NASA experts. And even if one did hit, damage would be slight and localized—not severe enough to endanger the community. One Princeton researcher has calculated that a hole in the skin the size of a window opening would incur a leakdown (depressurizing) time of 300 years!

Aside from all the obvious advantages of space colonization, the greatest impact in the end, perhaps, may be psychological—when people realize they are no longer on a doomed planet. With positive attitudes and hope for humanity restored, new horizons for progress on Earth may be as sweeping as space itself.

Application 9–9

Go back and skim-read "Cities in the Sky" to determine why you enjoyed reading it. Jot down at least two ideas that especially appealed to you.

Application 9–10

Study-read "Cities in the Sky" and answer the following questions:

1. What are some things that zero gravity will enable earthlings to do?
2. How will L_4 and L_5 be built and from where will most of the building materials come?
3. List some ways people remaining on earth will benefit.
4. Are there any problems on L_4 or L_5?
5. Why is agriculture placed in external pods?
6. What other things are done in external pods?
7. How is travel to be done between colonies?

The reader is not always at fault for inefficient reading. Often, a reader can't read well because the writer can't write well. You can't expect even a good driver to make good time over bad roads. A good reader learns to make progress through almost any kind of writing by adjusting the reading method to the particular situation. Good reading is not just step-by-step plodding, from left to right; it is fluid, flexible, easily adapted to the kind and quality of the material being read.

We have noted that knowing how to write better should help you read better. When you know how to develop ideas, you will recognize the basic, general, and supporting forms used by all writers, and you will know what the writer is trying to do before you read all of a paragraph or section. You can shift then from study-reading to skim-reading, or even to scan-

Reading better from writing better

reading, through that section. Remember that a piece of writing, no matter how long, contains one main idea, and all the other words and sentences contain ideas that amplify and adapt the main idea to the reader.

The article below about diving to the *Andrea Doria* will give you practice in adapting your reading to the type of writing and the author's purpose. Now that you know how a writer develops and organizes ideas, you can choose to scan, skim, or study, depending upon how much you need to concentrate in order to understand and respond as the writer intends. When you recognize the various kinds of developing forms and know why the writer is using them, you should be able to read with more understanding.

Application 9–11

In the *Andrea Doria* article, the words below are used in a special way, and their meanings must be derived from the way they are used in their context. Write the words on a sheet of paper, and opposite each write how the writer intended it to be understood.

secure into	plaque
habitat	umbilical
monitor	narcosis
excursion	vintage
decompress	ascend

Application 9–12

1. Study-read the following numbered paragraphs in the *Andrea Doria* report. Opposite each number print *D* if the paragraph is mainly description, *N* if mainly narration, or *C* if mainly causation.

_____ 8.	_____ 13.	_____ 14.	_____ 21.	_____ 24.
_____ 26.	_____ 28.	_____ 33.	_____ 37.	

Application 9–13

Read the following list of questions and proceed to study-read the deepsea diving report, searching for the answers. Do not write the answers until you have read the complete report. Doing this should help you to understand how important study-reading is to remembering what you read.

1. How many people were a part of the crew?
2. Did any other people go along with the crew?
3. How many divers could descend at one time?
4. How deep were the working dives?
5. What mainly hindered the divers?
6. When did cutting operations begin?
7. What is a "habitat"?
8. Did sharks hinder operations?
9. What kind of diving system was used?
10. What caused the expedition to end early?
11. How much did the expedition cost?
12. Was the expedition profitable?

Application 9–14 Write a summary of the main ideas in the *Andrea Doria* report. Also, state any conclusions you may have arrived at from reading it.

TREASURE HUNTERS DIVE TO THE ANDREA DORIA [5]

"How would you like to go to the *Doria*, Jack?" The question came from Bob Hollis, a diver friend and producer of documentary films.

"My bags are packed," I answered. I had been on one expedition to the *Andrea Doria* back in 1968, when an unsuccessful salvage attempt was made. Over a period of three weeks, I had made but three dives. I was eager to try again.

Bob explained that two young men in San Diego had founded a firm, called Saturation Systems, Inc., and were planning to use a diving habitat in a new salvage effort.

Both partners, Bob said, had impressive diving backgrounds. Twenty-seven-year-old Don Rodocker was a Navy-trained diver and expert mechanic, having designed a heat-exchange system used on two record-setting dives, one to 1010 feet. He also had served as a consultant to Oceaneering International, Inc., one of the largest deep-diving companies.

Chris DeLucchi was only 22 but he'd already accomplished more than most men do in a lifetime. When he was only 13 he installed an underwater platform—a kind of mini-habitat—in Monterey Bay, and at 15 designed and helped to manufacture underwater photographic equipment. Later, in the Navy, Chris became a qualified diving supervisor and, in 1972, established a new world's record—945 feet—for a saturation dive in the open sea.

Rodocker and DeLucchi had the qualifications, all right, but it remained to be seen if they could finally unlock the *Andrea Doria* treasure chest.

A number of expeditions had attempted and failed. Some had brought up items of token value, but the majority of the treasure seekers had been forced to abandon the project because of weather and rough diving conditions.

There remained aboard the ship valuable sculptures, silver wall plaques and jewels. (The *Andrea Doria* had boasted several jewelry shops.) There were vintage wines, probably still in good condition, and an estimated $1,100,000 in American and Italian currencies. No one had reached the first-class purser's safe since the sinking.

Bob Hollis, I and the other members of the photo team—John Clark, Tim Kelly and Bernie Campoli—flew from San Francisco on July 15, 1973, arriving 17 hours later at the seaport of Fairhaven, Mass., which was to be our base of operations.

There were 26 people on the expedition. In addition to back-up divers, boat crew and diving technicians, the group also included Sue DeLucchi and Barbara Rodocker, wives of the two leaders. George Powell, a Navy master diver, was in charge of the saturation diving system.

For deep diving, a mixed-gas atmosphere—oxygen and helium—must be used. As a general rule, diving with compressed air is limited to about 200 feet. At greater depths, the diver is apt to suffer nitrogen narcosis, which affects his ability to work and may even endanger his life. Helium has no narcotic effect.

To be done efficiently, deep diving calls for a saturation system. This allows the diver ample bottom time and eliminates a long decompression after each dive. He rests in a compression chamber in between dives and only decompresses (returns to surface air pressure) after his diving stint of days or weeks is over. Slow decompression is necessary to keep a man from being hit by the diving disability called the "bends."

The term "saturation" refers to the diver's tissues, which become saturated with an inert gas such as helium after he has spent a while in pressurized mixed-gas atmosphere.

"Mother," our nickname for the habitat, was the major component of the saturation system designed by Chris and Don. A steel cylinder 12½ feet long and 5 in diameter, it would be lowered to the *Doria* and tethered.

The habitat is capable of supporting three divers at a maximum depth of 600 feet for 21

---

[5] Jack McKenney, "Treasure Hunters Dive to the *Andrea Dorea*," *Popular Mechanics*, January 1975, pp. 72–75, 119, 120. Reprinted by permission of *Popular Mechanics*. Copyright © 1975 The Hearst Corporation.

days. An entry hatch at the bottom provides an exit to the sea.

An umbilical from the support ship on top can supply hot saltwater—to warm the divers' suits—electrical power, breathing and metal-cutting gases. There are audio and video communications systems linking Mother with the support ship. If all connections with the surface were severed, the habitat would be self-sustaining for at least five days.

Of necessity, life in the habitat is rugged, the only comforts being two fold-down bunks, a toilet, a food warmer and lights.

Diving operations aren't known for their creature comforts. Our support vessel, though a fine ship, proved no exception. The *Narragansett* was a fishing trawler, not a diving tender.

Just about every square foot of usable deck space on the *Narragansett* was occupied. The MCC (main control center), a 7-by-8-foot van, housed electronic and valving equipment to monitor the saturation dive from topside. When Chris and Don were saturated there would be a man stationed in the van at all times to regulate and check on the gas flow, heating and power, and to remain in communication with the divers.

The deck also held a big shark cage, a compressor system for scrubbing used gas from the divers, a welding unit and a boiler to supply the divers with hot water. Also taking up a considerable amount of space was a decompression chamber.

During the week that we were preparing dockside, people would come down to the dock to see what was going on. One time a little old lady walked up to Bernie Campoli and George Powell during an infrequent work break and said to George, "Do you think you can get my sister's rosary beads when you're down there on the ship? She was on the *Andrea Doria* when it sank, you know, and it would be so nice if she could get them back. She was in cabin 358 and they're in the drawer by her bed."

We left Fairhaven on Sunday, July 22, with Mother in tow behind the *Narragansett* and we arrived over the *Doria* the following day. It took us about six hours to locate the wreck and throw a grappling hook in. When it snagged, Tom Ingersoll, an exNavy salvage diving officer, and I made the first descent to identify and secure into the ship.

We dove on compressed air, which limited us to about 200 feet. At about 140 feet we expected to see the wreck come into view, but at 160 feet all we saw was a free swinging grappling hook. We had lost our contact with the *Doria*. Late that same day divers Don Gay and Tim Kelly managed to secure into the wreck.

This experience was typical of the things we would run into. Problems with equipment delayed us several times. We finally had Mother properly moored on the *Doria* on Aug. 7.

The diving proved formidable. It was deep—all of our working dives were between 150 and 200 feet, and in some cases dives were made to 230 feet. Currents were treacherous and visibility was usually limited to 25 feet. Everyone was dog-tired from hauling cables and umbilical lines, and working 12- and 15-hour shifts. More than once everyone was brusquely awakened at 1:00 or 2:00 A.M. during a tide change to prevent the umbilical line to Mother from tangling up in the anchor line. More hauling, pulling and lifting.

We were constantly aware of blue sharks sunning themselves on the surface and a number of times, while we were working or decompressing near the surface, they'd swim to within five or six feet of us. One time I looked back over my shoulder when I got a tingling sensation at the back of my neck. Not more than three feet away a seven-foot blue was turning.

While Tim Kelly and John Clark were in the water photographing Mother at about the 30-foot level, a big blue came in and started to harass them. John was using a camera with a large dome port on it. Apparently the blue was attracted to it because the shark came in and mouthed the camera while it was in John's hands. Shortly after that, Timmy literally had to kick one of the sharks off when it advanced.

Swift underwater currents made diving on the *Doria* difficult—and frightening. On one excursion down to the promenade deck there was only a hint of a current flowing south over the ship. We dropped down through the opening and spent about five minutes shooting pictures.

When it was time to surface, we popped our heads up out of the promenade area. To our surprise, a rope that we had secured loosely to the hull was suddenly picked up from the deck by an invisible hand and pulled right past us. We had to fight that strong current as we ascended the anchor line.

Figuring from the time of departure from San Francisco, it was almost a month before Don Rodocker and Chris DeLucchi were ready to begin cutting operations on the *Doria's* hull. It took another five days to cut a four-foot-square hole.

The two divers used Mother as their base and were connected to the habitat by long umbilicals that provided the helium-oxygen breathing mixture and communications with topside. Generally, one diver would make an excursion while the other, stationed in or near the habitat, acted as his tender.

Here is how Don Rodocker reported his first venture inside the *Doria*:

"On my descent I could see the tiled deck. Most of the tiles had fallen off and there were only a few remaining over the under-covering. I made a 10-foot excursion to either side; it looked like a disaster area and pretty dangerous to go poking around in the stuff. The water was fairly clear inside, but there was a lot of sediment on everything.

"I continued on down, playing my light over the rubble, and discovered a room still intact —also a set of stairs. The walls of this room must have been made of metal. It was the jewelry shop! I got to within 10 feet of it and could see mud circled in the corners of the windows, like snow on glass panes in the winter time. I couldn't see a thing on the inside.

"But immediately above me was a pile of debris precariously supported by only a couple of small timbers which had fallen and propped themselves against the wall. I realized that if my umbilical brushed against any of it, it could have collapsed and buried me. I couldn't get any closer than 10 feet."

The possibility that the interior of the ship was in bad shape had been discussed. But no one had imagined that it would be the scene of devastation we found.

Chris and Don had studied other liners and had found interior walls were made of metal with wood paneling on either side. (The metal would tend to contain a fire.) That was a standard method of construction.

But a new method had been used in the *Doria*. Her walls consisted of a nonmetallic insulation sandwiched between two wooden wall panels. When the wood disintegrated, nothing was left.

Chris and Don had anticipated salvaging at the least $2 million from the *Doria*, but that would have been possible only if the interior of the ship had been in reasonably good condition. The handwriting was clearly on the wall. The entire expedition had been at sea for 23 days. We had two scheduled days left, but with a report of bad weather heading our way, George Powell wisely decided to bring Mother to the surface and call an end to the expedition. It just wasn't worth risking the divers' lives for the little they might be able to salvage in the remaining two days.

In all, very little was salvaged. The most valuable find was an intact bottle of French perfume. But I don't think it will command a price of $150,000—approximate cost of the expedition. Diving is a risky business in more ways than one.

Application 9–15 After reading the next article, first tell whether description, narration, or causation is the main general form used to develop it. Next, copy the examples of definition, comparison, contrast, and analogy used to develop supporting paragraphs.

ORIGIN OF FIRE WHIRLS [6]

In meteorological terms, a fire whirl is a vortex—a fluid or gas mass with rotational motion. Vortexes are common in nature. Water flowing down the drain often develops a vortex. The familiar smoke ring is a vortex in air. A tropical hurricane is a vortex on a larger scale.

In general, a fire whirl is a stronger and more intense dust devil—in characteristics and mode of development the two are much alike. Both

[6] U.S. Department of Agriculture, Forest Service, *Fire Whirls . . . Why, When, and Where*, 1972, pp. 2–4.

depend on a supply of heated, buoyant air. On hot days when the skies are clear and winds light, dry land surfaces can become strongly heated by the sun—surface temperatures of 165°F are not uncommon. Under these conditions, a layer of superheated air is formed near the surface. This hot, buoyant surface air tends to rise in columns, which draw in more of the hot surface air as they move along with the wind. Some of the columns develop a strong rotational motion—these pick up dust and debris and become visible as a dust devil.

A layer of hot air overlain by cooler air is an unstable condition, and this in itself can start the updraft columns. More often, however, some sort of "push" is required to start the air moving upward. This may be created simply by wind flowing over some obstacle, such as a rock, small mound, or hill. In a fire area, the air is much more strongly heated than it can be over a surface heated by the sun—air temperatures exceeding 2,000°F in some parts of the fire are quite possible. Because of the intense heating, updraft columns start almost immediately when the fire is ignited. Except for very small fires, the lower part of the fire convection column is not a homogeneous mass of rising gas, but actually is made up of many separately rising columns—similar to the columns over a sun-heated surface. Like them, also, some of the fire columns develop rotational motion, and a fire whirl is born. Why some of the columns develop rotational motion whereas most do not is not yet clearly understood. Probably some sort of mechanical action, such as friction with obstructions, is required to start the air rotating. Once started, the rotation is intensified by the upward moving, buoyant air.

The airflow pattern in and around vortexes has some singular characteristics that give fire whirls their marked effect on fire behavior. In the central core of a well-developed whirl—the most visible part—the air movement is likely to be downward. Adjacent to the core, however, is a strong updraft. Speed of the rotating air outside the core is greatest close to the core, and decreases as the distance from the core becomes greater. It is the rapidly rotating air and the updraft that permit the fire whirl to pick up burning material and other debris.

The rapidly moving air and effects of centrifugal force tend to prevent air from entering the vortex from the side. Most of the air must enter near the ground surface where the rotational flow is slowed by friction. Thus, a relatively thin layer of horizontally moving air flowing into the vortex from all sides is created. The size and strength of the fire whirl determines how far from the vortex the airflow—and hence the fire—will be affected. Moderately strong fire whirls about 50 feet in diameter have been observed to affect the horizontal flow up to 500 feet from the whirl itself.

In artificially created fire whirls in the laboratory, it was found that the rate of burning of liquid fuels increased to 5 or 6 times the rate in still air. Although there is little experimental data about the amount of increase for woody fuels, observation indicates it is probably substantial. For woody fuels the increased rate appears to be largely due to the higher windspeeds, both in the whirl itself and in the inflow into the whirl. Since faster burning rate means greater fire intensity, the fire can be expected to spread more rapidly and be more difficult to control when whirls are prevalent. The increase in fire intensity also makes conditions more favorable for the development of additional fire whirls, so a chain reaction is set up—in more whirls, increased intensity, and still more whirls.

The rapidly moving air in a fire whirl, combined with its updraft, can tear loose and pick up burning material and inject it into the fire convection column. There it can be carried farther aloft and ahead of the fire. Thus, when whirls develop, there are more firebrands, and these tend to be larger and hence burn longer, so that they travel farther before they burn out. In southern California chaparral, fire whirls have been observed to uproot and carry aloft whole burning shrubs. In fires with fire whirls, there is likely to be more than usual long- and intermediate-distance spotting.

Fire whirls sometimes remain more or less stationary, but more often move in the direction of the prevailing wind or fire spread. When they move out of the fire area they usually lose their active fire, but scatter their firebrands into unburned fuel a short distance ahead of the fire or across the fireline. The many resulting spot fires can create an intense fire front very quickly.

In some fuels, spot fires start easily and build up rapidly; then the active fire in a whirl that moves out of the main fire is occasionally maintained as the whirl moves cross-country. When this happens, the fire is spread very rapidly on a narrow front, because the whirl can move with about the speed of the prevailing wind—ordinarily fire spread is much slower than the wind speed. This event is most likely to occur on relatively flat terrain, since individual whirls tend to dissipate rather quickly on moving out of a fire in rough country.

The ability of fire whirls to spread fire rapidly is a constant threat to fire crews. A single whirl can fan back to life even a fire that seems to be nearing control; it can scatter fire extensively across the fireline. When a whirl that moves out of the fire area contains fire, the hazard is obvious. But fire whirls no longer containing fire can also be dangerous. For a short time after moving out of the fire, they contain embers and hot gases, and dust and debris moving at high speed remain in the whirls through most of their life. The whirls may contain carbon monoxide and carbon dioxide, and may frequently be deficient in oxygen. Thus, fireline personnel should avoid whirls, even small ones.

Application 9–16 First skim through the following article to determine the purpose of the writer: to inform or to persuade. Next, skim through again, sifting out the main ideas supporting the writer's purpose, and just list them.

A NEW ENGINE CAN SAVE OLD BETSY [7]

Would you put $500 to $700 in your old car to get another 125,000 to 150,000 miles out of it?

That's what it would cost to replace the engine.

You can put new life in an old car for a lot less than $500. For something around $200—less than that if you do the work yourself—you can buy new tires, a tune-up, have the brakes relined and take care of any minor mechanical repairs. But $200 will buy only an additional 20,000 to 30,000 miles. A new engine will buy as many miles as you got out of the car when it was new.

A new engine is a pretty good buy when you look at prices of new cars. Any new car will set you back a minimum of $3000. A replacement engine will give you as many miles of transportation as a new car for a fraction of the price of a new car.

But there's another side to the coin. If your car has reached an age where the engine is shot, the car is probably showing its age elsewhere. Body rust, for example. Even if you replace the engine, you will still need tires, maybe a brake job and a general fixing up. But these items are considered routine maintenance and you would have to pay for them, new engine or not.

New engine or new car?

If you add the price of a new engine to what you would receive for your old car on a trade-in, you will probably have enough to make the down payment on a new car. You will still have to make payments for X number of months. As you build equity and the new car becomes an old car, you can trade in three or four years for another new car.

If you go for a new engine instead of a new car, the initial outlay will be much smaller, you won't be making payments for the next three years or longer and you should save on insurance premiums. On the minus side, an old car with a new engine won't bring much more than an old car with an old engine if you decide to trade in a few years. Used cars sell largely on appearance, not on mechanical condition.

But if you really like your present car and plan to keep it another four or five years, a new engine is a good investment.

There are several sources of supply for replacement engines. You can buy an original equipment engine—same everything as in a new car—from the auto companies. These engines are sold through automobile dealers. Detroit doesn't sell direct to individuals. You can also buy from the big mail-order houses, Sears and Wards, from jobbers who sell automotive parts, engine rebuilders and service outlets all of which specialize in replacement engines.

There are three types of replacement en-

---

[7] Robert Lund, "A New Engine Can Save Old Betsy," *Popular Mechanics*, June 1975, pp. 74–76. Reprinted by permission of *Popular Mechanics*, Copyright © 1975 The Hearst Corporation.

gines: factory originals, remanufactured and custom rebuilds.

Don't let the word remanufactured scare you and don't be misled by Detroit propaganda you may have heard about the dangers of using other than "genuine factory parts" in your car.

A remanufactured engine consists of a core that is put on a production line and rebuilt to original equipment specifications. The reputable remanufacturers are actually licensed by the car factories. That is, the car-makers tell the remanufacturers what they must do to bring the engines up to original equipment specs. All engines sold by mail-order houses are production-line rebuilds.

The third type of replacement engine is a custom rebuild. Replacement is a misnomer because the engine isn't replaced. It's rebuilt. A mechanic pulls the original engine out of your car, rebulids it, then puts the same engine back in. Most custom rebuilds are turned out by parts wholesalers and jobbers who operate rebuilding shops. There are some 6,000 to 7,000 custom rebuilders around the country.

If you want to do your own custom rebuilding, you can buy a kit with necessary parts from an auto-supply store or mail-order house.

The replacement engine business has a lingo of its own and it's important to know the language before you buy an engine or tackle the job yourself. If you don't know the language, you can be swindled or make trouble for yourself. Such as dismantling the engine—if you do the work yourself—and then discovering you don't have all the parts to complete the job.

There are different types of replacement engine packages within the three categories—original factory engine, remanufactured and custom rebuilt. The four basic packages are:

Package A, also known as a complete engine assembly. This is the best, most expensive, most complete replacement engine you can buy. You get about 250 new parts.

Package B, also known as a block assembly. You don't get as many parts as in an A package and some of the replacement parts are remanufactured, instead of new parts as in an A package. The block assembly does not include cylinder heads, rocker arms, oil pan or front cover, which are salvaged from the old engine.

Package C, also known as a short block, next step down from the B package. Fewer parts than in a B. The C deal does not include cylinder heads, oil pump, pick-up pipe, lifters, front cover, pully or installation gaskets. Again, these parts are salvaged from the old engine and used on the replacement.

Package D, the rebuild-it-yourself kit. Different kits contain different items. The Sears kit contains pistons, pins, rings, pin bushings, a reconditioned camshaft with bearings, oil pump, hydraulic or mechanical lifters, timing chain or camshaft gear and crankshaft gear, connecting-rod bearings and all seals and gaskets.

How do you decide among A, B, C and D?

One way is by price. A is the most expensive. B costs less than A. C costs less than B. D is rock bottom.

If you're considering a replacement engine, chances are you have taken care of your car. If that's the case you can rule out the A package. You don't need 250 new parts. Many of the original equipment parts are still usable. This narrows the choice to a B or C package if you are going to hire the work done.

Scott Connor, a consultant to the Automotive Parts and Accessories Assn., says replacement shops prefer to sell the bigger packages, the A or B, rather than the short block C. This isn't necessarily a case of oversell, of pushing more parts than the customer really needs, Scott explains. "The more precision work you take out of the hands of the mechanic, the less likelihood there is of making a mistake, of something going wrong."

Prices of replacement engines vary all over the lot, depending on the type of engine, what's included or left out of the package, the honesty/dishonesty and the competence/incompetence of the shop doing the work, make of car, whether the original engine was a Six or V8, the labor rate in your part of the country and a dozen other shakes of the dice.

If what you want out of a car is transportation, miles of use, and if you don't mind being a few years out of step on styling, a replacement engine is a good way to go. New car or replacement engine, you are buying about 150,000 miles of there-and-back. You get a lot more miles for the buck with a new engine than with a new car.

Application 9–17 Scan-read the preceding article by Robert Lund for at least two examples of classification used in developing it. Briefly describe them and tell what the standard is in each.

Application 9–18 Write a brief explanation of the way Lund arouses the reader's interest and motivates him or her to continue reading. Then tell how the writer adapts the ideas and the way they are expressed to the intended type of reader.

10

Purposeful business letters at work

memorandums . . . letters . . . their formats and main parts . . . main sections: (1) beginning, (2) developing and supporting, (3) concluding . . . kinds: (1) inquiry, (2) claim, (3) "Yes" and "No" replies . . . purpose: (1) to inform (2) to persuade

More people work today at producing words than at producing anything else. More money and time are spent on putting words together than on the construction of automobiles, airplanes, buildings, and highways all together. Much of that expenditure is for memorandums, letters, and reports. The essential cost of writing a single letter today is slightly more than six dollars, not including the cost of reading and responding to it. One technical report may cost thousands of dollars to prepare.

Unlike memorandums, which are intraorganizational, letters are usually directed to a receiver outside; they are interorganizational. Letters communicate the same kinds of ideas as memorandums, but usually are expressed in a more formal manner. This more formal tone reflects the writer's intention of being exact.

Reports identify and define an organization's important policies and decisions about its products or services. For this reason, reports are closely related to the survival of an organization, whether it be a manufacturer, a public hospital or other institution, or a labor union. Some of these reports contain the ideas without which a company cannot exist. Many describe scientific principles upon which copyrights and patent rights were established. Often they describe the results of years of expensive research—important discoveries that a company keeps secret, allowing it to compete successfully.

MEMORANDUMS

Memorandums are important in business because they enable easy communication between people in the same company. They also help to create goodwill between people on different levels of management and between management and other employees. They are often used between people who work together closely, who are acquainted and perhaps often have lunch together. Consequently, the language in memorandums is usually much less formal than in letters.

Most of the exchange of information within a company takes place by telephone. It is not unusual, however, for a person to ask that the main points of the telephone conversation be confirmed by a memo. This is done for verification or clarification.

The form of a memorandum is completely different from that of a letter. If a company has no prepared forms, headings may be typed at the top of the page as shown below. These are standard headings appearing in one pattern or another on almost all memorandums.

NAME OF COMPANY

INTEROFFICE COMMUNICATION

TO: DATE:

FROM:

SUBJECT:

The person writing or dictating the memo seldom signs it. Sometimes the signature is replaced by the sender's initials, placed where the signature would appear in a letter. It is not unusual that the secretary be required to place his or her initials at the bottom-left.

Memorandums are convenient forms of communication. Usually conversational in tone, they are used for the everyday details of doing business.

They contain routine information and thus are not always filed for future reference. Here is a typical example:

PANDA MOTOR CO.
Northville, Michigan

TO: Al Harmon DATE: 11/5/77

FROM: Mary Plaski

SUBJECT: Job 5330

Get in touch with Mashuri and see if you can convince him that we have to have a material certificate for a 12″ cable, Heat Code M4XKC.

Figure 10.1 is a somewhat more formal memorandum on a special form. Many companies prefer to use memorandum forms especially designed for their purposes.

PANDA MOTOR CO.
NORTHVILLE, MICH.

To All Motion Picture and Video Production Personnel	At Photomedia Department

Subject: Overtime Meal Allowance

As of October 29 there is no longer a meal allowance for overtime in the Photomedia Department. Record overtime hours as actual hours worked. If you start work at 8:00 a.m. and worked overtime from 4:30 p.m. until 9:00 p.m., and went out for a half-hour supper, you would record only actual time worked on your time card.

Example: 8:00-8:30 = 4 hrs. overtime

Signed *W. Latta* W. Latta, Unit Supvr.	Dept. or Location Photomedia	Date 11/4/74

Figure 10.1
Memorandum on special form.

Occasionally, a memorandum may be a very brief but complete technical report. The following memorandum report contains conclusions and a recommendation for the solution of a problem. It is written in a formal tone and from an objective viewpoint because accuracy is essential and because it is addressed to an executive with whom a formal relationship is judged appropriate.

TO: A. E. Madisrini, Vice President DATE: 5/11/77

FROM: R. U. Sontrim, Chief Engineer

SUBJECT: Stamped impressions for marking tool and dies

Stamping tool and dies for the purpose of permanently imprinting identification marks and dimensions would reduce their service life, and may lead to premature failures that otherwise might not have occurred. Stamped impressions create sharp sectional changes and induce stress risers. Sharp sectional changes are undesirable in the design of a component because of the uneven distribution of forces created by its geometry. Stress risers are minute surface stress concentrations resulting from cracks, tears, marks, and other abnormalities. Data collected by numerous industrial studies have confirmed that the presence of stress risers in a component substantially reduces its service life. Since it would not be good engineering practice to intentionally introduce stress risers and sharp sectional changes, such as stamped impressions, I recommend other less destructive alternatives such as painting, labels, etching, or electric engraving.

BUSINESS LETTERS

Letters are most often used for intercompany communications. Consequently, they are more formal in tone, because they must adhere at least to certain conventions of form and content.

You should try to vary the form of your letters to suit their nature. Because semiblock letters require that you indent between paragraphs, they are especially suitable for short letters that you want to double-space. They may be used also for single-spaced letters of medium length. The semiblock form is suitable because your indentation may use up some of the space in a double-spaced letter. In a single-spaced letter some space may be consumed not only by the indentation but also by the double space between paragraphs.

For a longer letter, the full-block form is better because it will provide more space. Full-block letters are usually single-spaced and no paragraph indentation is needed, so more sentences can be placed on the letter paper. To separate the paragraphs, however, you must double-space between them.

The accompanying diagrams of these two letter formats (Figure 10.2) show the most common parts of each and the spacing between them.

Business letters do not always have all of the items shown on the diagrams. Most of them, however, do have the following parts: (1) Letterhead or heading. A heading, inserted when the letter is not written on company stationery, consists only of the street address, city, and state. The name of the company, if there is one, is typed in capitals two spaces below the complimentary close. (2) Date. (3) Inside address. (4) Salutation.

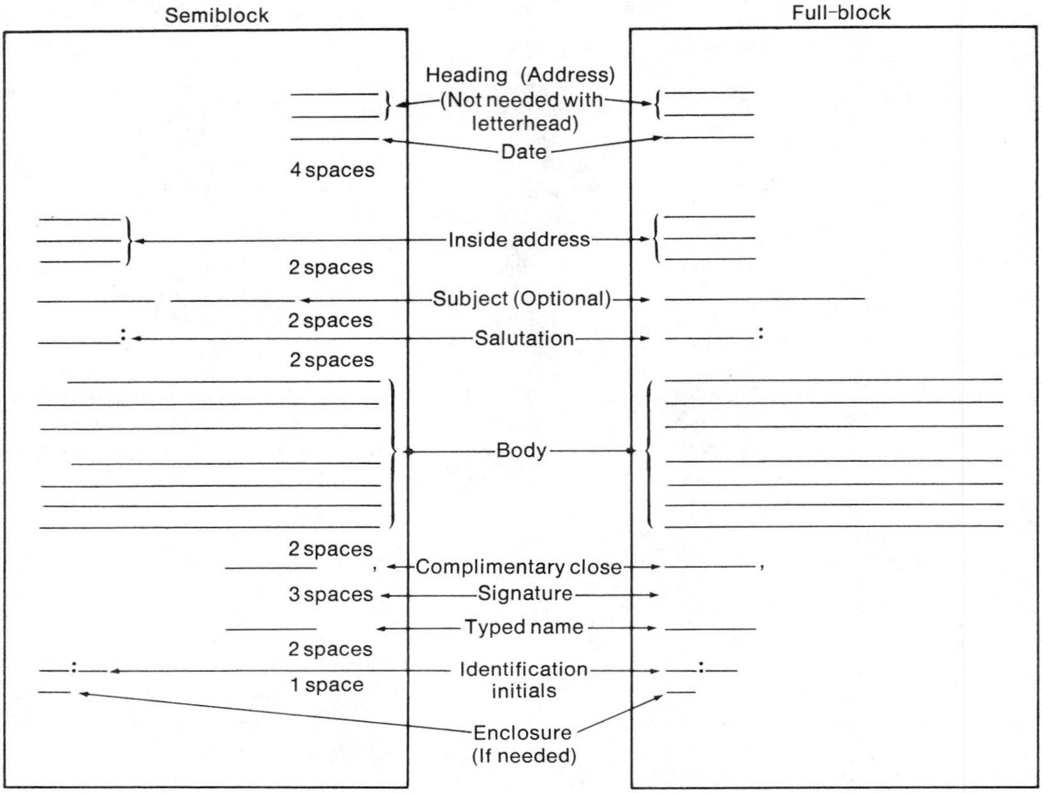

Figure 10.2
Letter formats.

(5) Body. (6) Complimentary close. (7) Company name. (8) Typed signature. (9) Handwritten signature.

The following additional components are optional:

1. *Subject line.* The writer may decide to use a subject line to save space. It lets him get directly to the main business without having to tell the reader what the letter is about in the first paragraph. It usually is introduced by the word "Subject:" or "SUBJECT:". The abbreviation "Re:" for "Referring to" or "Regarding" is also used for the same purpose. In either case, a brief, clear identification of the subject of the letter follows the colon.

   The subject line in a semiblock letter is indented five spaces and placed above the salutation, on a line opposite it. Most often, it is placed two lines above the salutation. In a full-block letter, it is usually placed flush against the left margin.
2. *Enclosures.* When enclosures are forwarded with a letter, their inclusion is indicated by the abbreviation "Encl." or "Enc." placed on the left side just below the identification initials. Sometimes the number of items enclosed is indicated in parentheses. When important documents or checks are enclosed, it is wise to list the enclosed

items—for example, "Enc. Check No. 540 to J. L. Long for $325.00." It is not unusual for the spelled-out word "Enclosure" to be used.

3. *Copy symbols.* The abbreviation "cc:" is placed two spaces beneath the enclosure line to indicate that copies of the letter have been sent. The names of the persons to whom these copies are sent may also be given.

Did you notice that this list does not include the "attention line"? Though it was once very popular, most writers now prefer to call attention to the person supposed to receive the letter by placing the name as the first item in the inside address, making it very conspicuous. This leaves room for another very important item, the "subject line."

The modified block and the simplified are two more often used letter formats. The modified block is a combination of the semiblock and full-block. Only its body is blocked, without paragraph indentation, but all the parts above and below it are, just as they are in the semiblock.

The simplified letter format is rapidly gaining in popularity because it greatly simplifies letter communication. Letters using this form are easier to type because they do not require all the conventions discussed in the others. Since they do not require a salutation, they eliminate the confusion about which to use. The indecision about whether to write "Dear Sir," "Dear Mrs. ______," "Dear Miss ______," or "Dear Ms." is avoided.

The simplified letter format below is intended for letterhead stationery. Otherwise, the heading would have to be inserted above the date, and the company name would have to be typed in below the typed name, as in the other forms.

Schoolcraft College

18600 HAGGERTY ROAD – LIVONIA, MICHIGAN 48151

April 27, 1977

Professor Dave Perkins
268 Building B
Wittenberg University
Bucyrus, Ohio 21403

SUBJECT: Simplified Letter

This example of a simplified letter format certainly would enable a person to avoid salutation confusion.

April 27, 1977

Professor Dave Perkins
268 Building B
Wittenberg University
Bucyrus, Ohio 21403

SUBJECT: Simplified Letter

This example of a simplified letter format certainly would enable a person to avoid salutation confusion.

This full-block character gives the letter a "let's get down to business" appearance. Its block form also makes it much easier to type.

The complimentary close is eliminated because it is no longer useful.

I urge that this form be used in all letters written for less formal communications.

Victoria Shiekh
Registrar

cc. Director of Admissions

Below is a full-block letter showing where the various parts are placed and the spacing between them. The beginning, developing, and concluding sections of the body vary in content, depending upon the kind of letter it is and its purpose. They will be discussed in detail later.

19330 Daly Road Detroit, Michigan 48240	Heading
November 10, 1974	Date
Mr. A. R. Sloan Sales Manager Cliff Corporation P. O. Box 9505 Toledo, Ohio 47736	Inside address
Subject: Wiring for Model 250 Voltmeter	Subject (may be in caps and indented)
Dear Sir:	Salutation
I recently purchased a Voltmeter, Model 250, and found the wiring instructions to be incomplete.	Body 1. Beginning section
The instructions were detailed to include cars and trucks built previous to the 1975 model year. My intent was to install this unit on a 1976 Toyota.	2. Developing and supporting section

Would you please send the correct wiring and any further information needed to install this unit on this particular model car.	3. Concluding section
Yours truly,	Complimentary close
W. R. BRADER COMPANY	Company name (capital letters)
Todd Markus	Signature
Todd Markus	Typed name
Field Service Engineer	Position
TM:tb	Identification initials
Enc. (2)	Enclosure
cc	Copies

Whether a letter is too short or too long is not determined by the number of words or sentences it contains. The length should be judged only by two things: (1) whether it contains enough words to achieve its intended purpose to inform or to persuade and (2) whether everything in it is related to what the writer wanted to say.

When a letter consists of more than one page, you should pay careful attention to the heading on the second and following pages. The heading of the second and subsequent ones must contain: (1) the full name of the recipient and right beneath it the name of the company for which he or she works, (2) the page number, and (3) the date. Here is the way it may look:

Mr. Henry H. Tomes, Agent Express Motor Freight Co.	—2—	March 14, 19__

The name of the company for which the writer works and his or her name are placed beneath the complimentary close. Below is an example showing the name of the company all in capital letters two spaces beneath the complimentary close.

Yours truly,

COMMERCIAL FORWARDING CO.

Todd Markus
Field Service Engineer

Kinds of Business Letters

Like technical reports, business letters may be of two general types, informative or interpretive. If a letter's main purpose is to inform or to request information, it is an informative letter. If its purpose is to persuade someone that a problem exists or to recommend that the receiver consider or apply your solution for one, it is an interpretive letter. The interpretive letter goes one step beyond the informative one; it communicates the writer's interpretations, what he or she concludes from the information. These interpretations are expressed as conclusions the writer draws logically from data. Knowing this difference will help you understand better what letters should contain.

More specifically, however, letters may be classified into several other common types. We shall discuss two main kinds in this chapter:

1. *Letters of inquiry and replies to them.* The main purpose of this type of letter is to inform, to provide the reader with information about products, services, costs, etc. One may be sent along with samples, pamphlets, and other kinds of literature. The letter of inquiry asks for information, and the reply responds to it.
2. *Claim and adjustment letters.* Claim letters, also called complaint letters, usually register the writer's unhappiness or dissatisfaction. They may be written to correct a problem or a misunderstanding related to an agreement, either stated or implied. An order arrives with the wrong quantity or quality of the item ordered, and the writer requests or demands an adjustment. It results mainly from a difference of interpretation of what was ordered or of what actually happened.

   The adjustment letter is the reply to the claim letter. It expresses the writer's interpretations of what happened. One expresses the reaction to the original claim letter, telling the claimant whether the claim will be honored or rejected partially or completely.

Letters of Inquiry and "Yes" and "No" Replies

One of the most common forms of business letters is the kind asking for information, that is, the inquiry letter. When you receive one, to answer it you may have to devote a great deal of time and effort to get the requested information. Therefore, when you write a letter of inquiry asking for information, you will understand why a variety of persuasive devices should be used to get the reply that you want. Your letter should develop the relationship needed for that purpose.

Below is the format for a letter of inquiry, showing the care the writer takes to get the right kind of relationship. It also shows the main sections your inquiry letters should have and the kind of details these sections should contain to enable the letter to do what it is supposed to. Although every inquiry letter does not have to have all of these items shown below, many letters contain most of them.

As stated earlier in this chapter, every letter can't be as neatly divided as shown below. A single sentence in one may perform two or three of the functions listed. Dividing letters into these main sections is done mainly to help you. Studying letter writing in this way will teach you how to focus your mind on one thing at a time. This will enable you to identify more exactly and to put into the letter the kinds of details needed to do the job you want it to do. Consequently, you will succeed in writing a more purposeful letter.

A. *Beginning Section*
1. Comment on the pleasant mutual relationship.
2. Clearly and accurately identify the question to be answered or the item requested.

B. *Developing and Supporting Section*
1. Give any needed definitions or background information.
2. Explain why you want the reply and how you intend to use it, if helpful.

C. *Concluding Section*
1. Indicate confidence that your request will be answered.
2. Offer to send any other needed information.

When writing the letter of inquiry it is important that you identify clearly, briefly, and exactly the request you want answered. When you don't, it is sometimes necessary for several exchanges of letters to take place before the person to whom you are writing can provide you with what you need.

Here is an example of a letter of inquiry with the question clearly stated. Like all the other letters used as illustration in this book, it is not on stationery; consequently, the name of the company is capitalized and placed two spaces beneath the complimentary close.

1250 Royal Drive
Plymouth, Michigan 48170
November 1, 1974

Mr. John Smertski
Power Products Distributors
33133 Eiffel Court
Warren, Michigan 48193

Dear Mr. Smertski:

For years I have been ordering chainsaw blades from you for my company. While looking through a farm equipment magazine, I noticed a new carboloid tipped blade and was wondering if your company carried such a product. I am very interested in a blade of this type and would appreciate some information and reports of test results. A–1, 2

It seems to me that the hard, high abrasive tungsten-carbide alloy will revolutionize the tree service business. With our men not stopping to B–2 change or sharpen blades, my company should save a great deal of money, not to mention the faster service we will be able to give the public.

I hope to be receiving your literature shortly. However, if you do not C–1
carry this product perhaps you can suggest someone who does. Your
assistance on this matter, as always, will be greatly appreciated. I have C–2
enjoyed dealing with you in the past and look forward to doing business
with you in the future.

Yours truly,

ASHBURN TREE SERVICE

Vladimir Drenko
Equipment Manager

VD:ra

Letters written to ask for credit often are merely letters of inquiry. Therefore, somewhat the same format as that used in writing inquiry letters may be used. In this type of letter, however, information more specifically related to credit standing must be given.

The following is a suggested format for a letter requesting that you or the company you represent be placed on an approved credit list. Remember, every credit application letter need not have all of the items listed under each heading below; however, all of them may be very helpful in securing credit.

A. *Beginning Section*

1. Refer to the past pleasant business relationship.
2. Request the kind of charge account you need, stating any special terms you require.

B. *Developing and Supporting Section*

1. Relate any needed background information, e.g., the kind of business, a brief history of it, a brief history of the relationship with this particular company.
2. Tell how you have been paying your bills in the past, C.O.D. or within ninety days, and explain why you did not apply for credit before.
3. Give some reliable references for the soundness of your credit, e.g., banks or the suppliers with whom you already have credit accounts in good standing.
4. Present one or two good reasons why you want to change from the present way of paying. Perhaps the increase in the volume of purchases make it impractical to keep much cash on hand.

C. *Concluding Section*

1. State that you are sure that your request will be given fair consideration.
2. Assure the receiver that you will comply with all of the credit terms.
3. Close with a statement that you look forward to a continued pleasant business relationship.

Below is an illustration of a credit inquiry letter following the preceding outline:

36236 Industrial Road
Centerville, Michigan 48001
February 15, 1975

Automatic Screw Machine Co.
1313 Lakeview Drive
Columbus, Ohio 30806

Gentlemen:

We are grateful for the wholesome business relationship that has A–1
been established between our two companies. Therefore, we feel sure
that you will honor our present request to be placed on your list of ap- A–2
proved credit accounts. As you know, we have a C.O.D. arrangement at
the present time.

Our company is now two years old, and during this time, the volume B–1
of brass and steel screws we manufacture has increased tremendously.
This, of course, requires that we purchase large amounts of materials B–2
and equipment for which we are now paying C.O.D.

Enclosed is a list of some of the companies with which we already
have established charge accounts along with a profit and loss statement B–3
for the last year. We will gladly provide additional credit references.

Before our increase in sales, our C.O.D. arrangement with you was
adequate, but now we know you will understand why we must change. B–4
It has become difficult for us to keep safely enough cash available at all
times to pay for large deliveries.

I am sure that you will give our request fair consideration. Knowing C–1
that you are sure that we will comply with all of the terms of your credit C–2
plan, we look forward to our continued pleasant business relationship. C–3

Yours truly,

DIEHARD MACHINE CO.

Clarence P. Thorndike
President

CPT:hp
Enc. (2)

'Yes" and "No" Replies to Letters of Inquiry

It is not unusual for the letter replying to an inquiry to require more care and effort than the original inquiry. These replies may be either of two kinds: (1) "Yes" reply, providing everything asked for, information, samples, etc., or (2) "No" reply, stating that the request will not or cannot be satisfied completely or partially.

Favorable replies, "Yes" letters, answer the questions in a friendly manner. You should avoid showing irritation even if the information was difficult to get or expensive. Even routine things like sending literature, a catalogue, or a form letter reply should be courteously done.

Following is a brief outline showing a suggested format of a "Yes" reply to a letter of inquiry:

A. *Beginning Section*

1. Refer to the past communication (spoken or written) in which the inquiry was made.
2. Restate the request exactly as you understand it, and enthusiastically give the "Yes" reply.

B. *Developing and Supporting Section*

1. If necessary, explain your answer further by:
   a. Providing background information.
   b. Giving causes and reasons related to your answers.
   c. Providing needed definitions.
2. Inform the inquirer of any precautions that should be taken in using the information to avoid injury, damages, or costly wastage of material.

C. *Concluding Section*

1. Identify any additional enclosed literature or things sent in reply.
2. Comment about the pleasant features of the business relationship and offer to be of further help.

Below is a good example of a "Yes" letter reply to a letter of inquiry following the format on the preceding page:

33133 Eiffel Court
Warren, Michigan 48193
November 6, 1974

Mr. Vladmir Drenko
Ashburn Tree Service
1250 Royal Drive
Plymouth, Michigan 48170

Dear Mr. Drenko:

In your letter of November 1, 1976, you inquire as to whether we can order a carboloid tipped blade for you. I am happy to tell you that I have checked with our supplier, and two of these blades are in stock. A–1 A–2

The supplier with whom I spoke about this blade assured me that you are definitely right about its being a time and money saver. After B–1
you have used it for some time, please let me know whether it is better than the blade you have been ordering from us. I do value your judg- B–2
ment and would like your evaluation so that I can pass it on to other customers who use this type of blade.

Since this blade is in short supply because it is new, I suggest that you purchase the two now in stock. This will enable you to have a spare so that work stoppage while waiting for a replacement for a broken blade can be avoided.

Enclosed are some pieces of literature about this blade. I hope I C–1
have been helpful. Please reply as soon as you can to tell me whether I should place your order for one or two blades. I'll be happy to try C–2
to give you any other information you may need.

Yours truly,

POWER PRODUCTS DISTRIBUTORS

John Smertski
Purchasing Department

JS:ab
Enc.

Unfavorable replies, "No" letters, especially those inquiring about credit, are much more difficult to write than the request letters. "No" letters are even more difficult when the goodwill of the receiver is to be preserved.

The following is an outline showing how a rejection or a refusal reply ("No" letter) to a letter of inquiry may be written. Remember, it doesn't have to have all of the suggested items listed below. However, it should have most of them.

A. *Beginning Section*

1. Refer to the oral or written communication in which the original inquiry was made.
2. Restate the request in your own words to show that you clearly understand what is wanted.
3. Briefly tell the reader what you had to do (time, effort, cost, etc.) in trying to honor the request.

B. *Developing and Supporting Section*

1. Express sincere appreciation of the problem.
2. Give the main reasons, without refusing the request, explaining why you can't grant it.
3. Now clearly state you can not honor all or part of it.

C. *Concluding Section*

1. Indicate your unhappiness in not being able to grant everything requested.
2. Express your desire to continue your business relationship, and be sure to point out the advantages to be derived from doing so.

Below is a full-block "No" letter stating that the writer is unable to provide the information requested about the carboloid blade.

33133 Eiffel Court
Warren, Michigan 48193
November 6, 1974

Mr. Vladmir Drenko
Ashburn Tree Service
1250 Royal Drive
Plymouth, Michigan 48170

Dear Mr. Drenko:

As soon as I received your letter of November 1, 1974, asking whether A–1
we can order a carboloid tipped blade, I began a thorough search. After A–2
checking our own inventory sheets, I went through a number of cata-
logues, but was unable to find it. I then made telephone calls to several A–3
suppliers, including some in distant states.

I know how important this blade is for the the kind of work you are B–1
now doing; that is why I spent so much time searching. I am sorry B–2
that I must tell you that we cannot get the carboloid blade you want. B–3

I hope that our inability to supply this particular blade will not curtail C–1
your other business transactions with us. Please do not hesitate to C–2
call me at any time.

Yours truly,

POWER PRODUCTS DISTRIBUTORS

John Smertski
Purchasing Department

JS:ab

It is much harder to write a reply to a credit inquiry telling the applicant that credit cannot be granted at this time. In the following outline and letter, notice how carefully the writer explains why credit must be refused before actually stating the refusal. This makes it easier for the receiver to accept the refusal, because the reasons justify it.

Although the letter need not contain all of the items shown below, it is usually more successful if most of them are included.

A. *Beginning Section*

1. Refer to the communication (oral or written) in which the request for credit was made.
2. Express appreciation for any past business and comment on the current pleasant business relationship.

B. *Developing and Supporting Section*

1. Explain how you went about investigating the credit standing of the applicant, showing that you made a thorough and fair investigation.
2. Give the reasons why you cannot extend more credit at this time. Be careful to avoid any loaded words, such as "*We consider you a poor risk.*"
3. State regretfully that you cannot extend the credit requested.
4. Suggest steps to enable the desired credit to be granted in the future.
5. Encourage the continuing of the present transactions by pointing out the advantages in doing so.
6. Give whatever suggestions you can to ease the burden of present payments. Suggest, perhaps, larger-volume orders at lower prices, to be paid in 90 days on a "same as cash basis."

C. *Concluding Section*

1. Indicate your unhappiness at not being able to extend the desired credit.
2. State that you will be happy to answer any other questions.
3. End on a sincere friendly note and, if you think it helpful, offer to have a representative go there to explain further.

Below is an unfavorable reply to a credit inquiry letter. Notice how the writer, before actually saying "No," carefully prepares the reader to accept the answer as a reasonable one.

1313 Lakeview Drive
Columbus, Ohio 30806
March 5, 1975

Mr. Clarence P. Thorndike
President
Diehard Screw Company
36236 Industrial Road
Centerville, Michigan 48001

Subject: Your letter of February 15, 1975 requesting credit arrangements.

Dear Mr. Thorndike:

I have not heard from you since our last meeting at the Machine
Tools Conference of 1974. I was happy to learn from your letter of A–1
the success you are having. We consider you one of our best cash A–2
accounts and are grateful for your business.

In investigating your credit standing, we were mainly concerned
with the degree of your indebtedness at the present time. We wanted
to know, especially, whether or not you still have enough unmortgaged
collateral. We found that your main customer is Ingersoll Office Machine Company. Other than Ingersoll, your customers are mainly dis- B–1
tributors who buy small quantities of screws for the replacement
market. We concluded that if Ingersoll suddenly decided to purchase B–2
its screws from one of several larger suppliers, the banks from whom
you borrowed for your new buildings would have to foreclose. In view
of our findings, we have to tell you that we can not grant you "Unlimited Credit" at this time. B–3

Until you are able to reduce the number of first mortgages on your
property and until you are able to get a few more large customers like B–4
Ingersoll Office Machine, I suggest that you take advantage of one of B–5
our credit plans with a limit, perhaps, $15,000. To keep you as a
customer, we are willing to offer you prices which we give only to our B–6
preferred large-volume credit customers. That will require you only to
place your order a year ahead of the delivery date. This enables us to
do work on your order during our slack seasons.

It is unpleasant for me to refuse the kind of credit terms a good C–1
customer wants. But at the rate you are growing, I am sure you will be
able to get "Unlimited" credit from us soon. I will be glad to answer C–2
any further questions you may have. I look forward to meeting you at C–3
the Machine Tools Conference next week to explain further.

Yours truly,

AUTOMATIC SCREW MACHINE CO.

Arthur L. Sipes
Treasurer

ALS/jr

Claim Letters

The main purpose of a claim letter is to secure a fair settlement for a claim or to register a complaint. The claim or dissatisfaction may be expressed about a late, a damaged, or an incomplete delivery of merchandise. It may also be related to the delivery of merchandise that was never ordered. The claim letter may call attention to the installation of faulty equipment or express dissatisfaction with services. Claim letters are often written to point out mistakes made in the preparation of invoices.

Regardless of which of the many claims or complaints a letter may express, you can't achieve your goal of securing a satisfactory adjustment unless you present your contentions in a courteous, persuasive manner. Your main purpose will be to persuade the reader to honor your claim; you will not be only informing. Therefore, attacking the person by calling names, by verbally shaking your fist at the reader, or by threatening in any other way will achieve very little. The person receiving your persuasive letter will be much more willing to go out of the normal procedure to help make the adjustment you want.

To help write a more effective claim letter, you should presume that the error made was accidental and unintentional, and that the person you are directing your claim letter to has little or no responsibility for the mishap until you have proof otherwise. For example, a piece of equipment shipped may have been in perfect condition when it left the manufacturer's warehouse. The billing clerk for the railroad, however, may have incorrectly prepared a shipping order, causing the shipment to end up in some distant city. The resulting delay in delivery, of course, was not the fault of the manufacturer.

Below are the main sections of a claim letter together with the ideas most often included.

A. *Beginning Section*

1. Identify exactly the transaction or agreement to which the claim is related: the correct date, order, invoice, or bill of lading number, accurate sums or totals of money, or other items involved.
2. Clearly state your complaint or claim, and, if possible, identify the terms, promises, guarantees, or laws violated.
3. State what you consider a fair adjustment.

B. *Developing and Supporting Section*

1. Describe more specifically the extent of your claim.
2. Identify and explain, if you can, what you think may have caused the problem.
3. Justify your claim by pointing out the consequences you suffered: loss of money, time, customers, production delays, etc.
4. Identify the action you reluctantly may have to take if your claim is not honored.

C. *Concluding Section*

1. If you are not sure who is responsible for what happened, state it.
2. Express confidence that the person receiving your letter will do everything possible to see that your claim is adjusted fairly.

When the relationship between the people involved in a claim does not require extensive explanations, a somewhat briefer one may be written. The following letter is an example.

143 S. Main
Plymouth, MI 48170
October 29, 1974

Good Furniture Company
564 Royal Road
Livonia, MI 48154

SUBJECT: Receipt of damaged chair, our Order No. 3546.

Gentlemen:

On October 25, 1974, I purchased a chair from your Euclid Road store for the sum of $169.95, Order No. 3546. The terms of the contract state, "Any damage during shipping will be replaced free of charge." A–1 A–2

The chair arrived with a long tear in the backrest fabric. Please send a new chair replacement. The chair is a wedding gift, and the customer is eager to receive it before the wedding. If you can't send this same model, send one with the same pattern even though it may be brown, green, or tan. Don't send any other colors or patterns. A–3

The tear in the chair's fabric is about seven inches long. The rubber padding beneath the fabric is cut to some extent. This leads us to think the damage may have been done by a lift truck at your warehouse or at the truck terminal. We hope you will take care of this matter so we can continue purchasing furniture from you, and so that we can regain the confidence of this unhappy customer. B–1 B–2 B–3, 4

I know these things happen during shipment, and this damage might not be your fault. My past experience with you assures me that you will handle this matter fairly. If you need any additional information, let me know. C–1 C–2

Sincerely,

ELMER'S QUALITY FURNITURE CO.

James Earl
Manager

JE:lc

Claim Adjustment Letters

The reply to a claim letter, often referred to as claim adjustment letter, may grant the claim or reject it, or sometimes offer a partial settlement. Obviously it is more pleasant to write a "Yes" than a "No" letter.

Here is a suggested way to write an adjustment letter agreeing to settle the claim in full:

A. *Beginning Section*

1. Refer to the past communication, conversation or letter, presenting the claim.
2. Identify the exact claim and the solutions or adjustments suggested by the claimant.
3. Acknowledge the justification of his claim and tell him pleasantly that you are granting it.

B. *Developing and Supporting Section*

1. Express regret for the inconveniences, damages, or financial losses suffered.
2. Briefly explain, if you can, the causes for the mishap.
3. Tell what is being done to make sure that the same thing does not happen again.

C. *Concluding Section*

1. Reinforce his confidence that you will do everything possible to keep and strengthen your relationship with him.
2. Close on a friendly note.

The example below of a "Yes" reply to a claim letter also illustrates how a single sentence may contain more than one of the items suggested in the preceding letter format. The first sentence in it has three of the items suggested under "The Beginning Section."

564 Royal Road
Livonia, Michigan 48154
November 19, 1974

Mr. James Earl
Elmer's Quality Furniture Co.
143 S. Main Street
Plymouth, Michigan 48154

SUBJECT: Damaged Chair on P. O. 3546.

Dear Mr. Earl:

In reference to the claim in your letter of October 29, 1974, concern- A–1
cerning the damaged chair on purchase order #3546, our bill of A–2
lading indicates that the damage was caused at our warehouse, and we
are happy to send you a new chair at once, without cost to you, as you A–3
requested.

We regret any inconvenience you may have suffered. This mishap B–1
would not have happened if our drivers had securely tied the chair
down. Orders have been given to make sure this is done on future B–2
shipments. Your chair is on its way. B–3

We hope this will not happen again, and we look forward to a con- C–1
tinuing pleasant relationship. We value your business. C–2

Sincerely yours,

GOOD FURNITURE COMPANY

Frank Lawrence
Supervisor

FL:lc

The ability to say "No" diplomatically is one of the most valuable skills a person in business can acquire. The ability to tactfully refuse a claim or complaint and still not lose a worthwhile customer or employee is often required. Because you won't be able to give a "Yes" answer to the person expecting one, you will start out with a distinct disadvantage in writing this kind of letter.

Most of the unhappiness, irritation, or anger a claimant may experience results from his or her inability to understand clearly the reasons for the rejection of the claim or the truth of the complaint. The main thing a refusal letter has to do is to say "No" in a way that will enable the reader to understand to some extent the fairness of your doing it.

To enable your readers to understand the justification of your refusal to honor the claim, you might explain how they will benefit by the care with which you consider all claims. You might point out that the company's policy of being very careful is essential if it is to keep offering their products or services at such low competitive prices.

If you can't honor a claim, do whatever you can to suggest some other way for the claimant to get satisfaction. If you think the railroad or truck line damaged the merchandise, suggest that either be contacted and claim be filed with the one responsible. You might suggest that an insurance company may be obligated to reimburse for the lost or damaged material.

Following is a suggested format for a refusal reply to a claim letter, a "No" letter. It is a good one to follow until you are able to improvise one of your own that does the difficult job as well.

A. *Beginning Section*

1. Acknowledge receiving the communication reporting the claim.
2. Assure the claimant that you understand the claim and restate it, especially the dates, amounts, and what the claimant wants in settlement, to show you both have the same things in mind.
3. Express your sympathy and unhappiness that it occurred.

B. *Developing and Supporting Section*

1. Report in detail the things you did, the time you spent, and the sincere effort you exerted in investigating the claimant's assertions.
2. Give the findings of your investigation, focusing especially on the reasons that will justify the conclusion you will give next.

3. State that you cannot honor the claim as a result of your findings or any other reasons, such as company regulations or policies.
4. Suggest others who might have been responsible for the mishap and offer any other solutions for the problem worthy of the claimant's investigation.

C. *Concluding Section*

1. State sympathetically that you understand the claimant's problem.
2. Assure him or her that you have done everything you are authorized to do to adjust the claim fairly.
3. Emphasize the advantages for the claimant in continuing the business arrangement and comment on the pleasant relationship you want continued.

Following is a "No" reply to a claim letter:

564 Royal Road
Livonia, Michigan 48154
November 14, 1974

Mr. James Earl
Elmer's Quality Furniture Co.
143 S. Main Street
Plymouth, Michigan 48154

Subject: Damaged Chair, Purchase Order #3546.

Dear Sir:

About your letter of October 29, 1974, reporting a claim on a damaged A–1
chair, purchase order #3546, which costs $196.95, I wish to express A–2
my regret that this chair arrived with the backrest torn. A–3

We devoted three days to investigating your report, and we found that
the chair was picked up at our warehouse by the railroad in good B–1
condition. Our company policy will not allow us to pay for the
damages caused by a carrier. Therefore, we sincerely regret that we B–2
must tell you we cannot honor your claim. We suggest you report B–3
your claim to the O.S.&D. clerk for the carrier delivering. B–4

We understand your claim and agree that this is an unfortunate C–1
happening. We have done everything possible to honor your claim;
however, since the carrier caused the damage, we feel sure that they C–2
will reimburse you for it. We want you to continue taking advantage C–3
of our low prices on furniture.

Sincerely,

GOOD FURNITURE COMPANY

Frank Lawrence
Supervisor

FL:lc

Purposeful business letters at work

Other types of technical letters will be discussed later. Employment application and résumé cover letters will be explained and illustrated in the next chapter.

Application 10–1

Read the claim letter below carefully and do the following:

1. Rewrite it so that it effectively achieves its purpose.
2. Write a reply stating you will honor the customer's claim, a "Yes" letter.
3. Finally, write a "No" letter, one that diplomatically says you cannot honor the claim.

Be sure to write a complete letter for each, showing all of the components, including the heading and typed signature. Also, be sure to follow the suggested outline for each kind of claim letter and include as many as you can of the items listed in each suggested format. Don't forget to correct any errors you detect in spelling and punctuation.

Gentlemen:

Three weeks ago I ordered and had installed a hot water tank from your local distributor in Newport News, Virginia, The Plimpton Plumbing Supply Company.

You are the manufacturer and I want to say this gadget just won't work right. It's heating unit isn't large enough, I think. Last week I got mad at it and told my brother to try to fix it but he couldn't and I wasn't able to get my housework done.

What's the matter with all you guys, can't you build anything right. If you don't get me another tank right away I'm going to tell my lawyer to sue you.

I'm warnin you now,

S. Omar

Application 10–2

Write a letter presenting your inquiry about something in the field in which you are now involved either at college or on the job—for example, gas station, supermarket, engineering, accounting, etc. If you prefer, write about a field in which you hope to be involved sometime in the future. In the right margin opposite each letter, insert the letter of the alphabet and number to identify each section and subsection, as in the examples.

Next, write a letter answering the inquiry. Now write another telling that you cannot give the answer and why you can't give it.

Application 10–3

Do the same as the above for a credit inquiry letter.

Application 10–4

Imagine that you are the owner of a small grocery store. Write a full-block letter asking a wholesaler about the prices on a product.

Application 10–5

Write a semiblock letter telling a railroad that one of the packages in a shipment arrived damaged. To do this, imagine that you are the owner of a small business—a restaurant, garage, or the like.

Application 10–6 Identify each of the following letters as to type (claim, "Yes" adjustment, etc.). Identify the sections according to the model formats illustrated in this chapter, by writing the appropriate capital letter and number.

A. Dear Mr. _____:

This is an answer to your inquiry of November 5, 1976, requesting information on the XY-600 copy machine. I am very appreciative of Mr. Fadden's recommendation to you, and I want you to know that I will be glad to help you in any way I can to solve your copy-center problems.

The XY-600 is a very fine copy machine; however, I feel that if your office will be making 2000–3000 copies a month, you may need a slightly heavier one. Therefore, in addition to the enclosed material that I am sending you describing the new XY-800, I have asked Mr. Major to send you materials from his division, related to the cost and maintenance of this machine. I am sure that you will find all this material interesting and informative.

I have talked with your secretary and have set a meeting time on Wednesday, November 20th, at 1 o'clock to get together and analyze your office situation. I will do everything I can to help correct and improve this copy-center.

Thank you again for your confidence in me. Please feel free to call me if I can be of further assistance before our meeting on the 20th.

Yours truly,

B. Dear Mrs. _____:

In response to your letter dated October 28, 1977, I must apologize for the delay of your household goods shipment and for any inconvenience it may have caused you. I can assure you that every effort is being made to provide you with the exact date on which you can expect to receive your shipment.

I can certainly understand the problems created by this delay, and I hope to rectify this unfortunate situation as soon as possible. At the present time, I cannot give you a satisfactory answer. My inquiries have revealed that your shipment hasn't arrived at its United States port of entry. Until it does, there is very little I can do. I have, however, requested the Naval Supply Depot in Yokosuka, Japan, to initiate a tracer on your shipment and to forward the results to me immediately. With that information, I can at least provide you with a rough estimate of when your shipment will arrive in Detroit.

I am sorry this reply will not meet your present needs, but it is the best I was able to do. Any further information I can give you will be relayed to you by telephone. If any other questions occur to you, please call collect.

Yours truly,

C. Dear Ms. _____:

I would appreciate any information you can give me on the SY 600 copy machine. Mr. Fadden, our Purchasing Agent, suggested that I contact you. He assured me that you would give me the same excellent advice and service you have been giving him.

In our office, we would be running approximately 2000–3000 copies of miscellaneous material each month. Do you think the XY 600 is the ideal machine for this amount of use? Perhaps there is another model that would be more efficient in our office situation. We would appreciate your advice and comments.

Please call my office and set up an appointment with my secretary as soon as possible so that we may discuss this matter further. I look forward to meeting you.

Yours truly,

11

Writing for employment

letters of application . . . résumés . . . cover letters . . . what they are . . . their main sections . . . how to prepare them . . . their purposes . . . the employment interview

A type of business or technical letter everyone should learn to write correctly and effectively is the employment application letter. This may be the most important kind that you will ever write, unless you are hired by a relative and remain on that job the rest of your life. Even if you become self-employed, you will have to know enough about employment letters to enable you to judge the values of those who write to you for a job.

There are two main kinds of employment application letters. Both are written after you know that an opening actually exists. They are not letters asking whether or not there is a vacancy, just another kind of letter of inquiry. These letters are the (1) employment application and (2) résumé cover letter. The résumé cover letter is written after the résumé, and it is sent along with it.

Securing a wanted job is the main purpose of both of these letters. Usually, however, an interview is required before a person is hired. The main immediate purpose of both of these letters, therefore, is to persuade the reader that you are worth an interview. They do this by pointing out clearly the qualifications making you especially suitable for that particular opening.

EMPLOYMENT APPLICATION LETTER

The letter of employment application is easier to write than the résumé because it consists of only the letter, not the supporting data sheets. Its main functions are:

1. To tell how you learned about the opening.
2. To identify the specific job.
3. To state briefly and clearly the related educational and occupational experience.
4. To give needed personal data.
5. To list one or two references.
6. To request an interview and indicate how you can be contacted for scheduling it.

You can see how the preceding suggested details are included in the following employment application letter format:

A. *Beginning Section*

1. Identify which position or which type of employment you want.
2. State how you learned about the opening: newspaper ad, employment agency, a friend or relative.
3. Tell why you are interested in working for that particular company.

B. *Developing and Supporting Section*

1. Clearly point out your strong qualifications, emphasizing those that make you a good choice for the opening.
2. Identify employment you have had, especially that related to the opening.
3. Briefly summarize your education, emphasizing the training related to the opening.

C. *Concluding Section*

1. Request an interview at the receiver's convenience. If you are applying for a job in a distant city, tell when will be the best time for you to go there for the interview.
2. Clearly tell where you may be reached, including your telephone number.
3. Tell whether or not he or she may contact your present employer for reference.
4. Offer to send any other needed information.

Notice how the writer of the following letter of application gives and suggests some good reasons why he would be worth considering for the position.

Employment application letter

1267 Theota Avenue
Ann Arbor, MI 48935
November 5, 1975

Superintendent of Schools
Parma Public Schools
Personnel Director
Administration Building
1198 W. 54th Street
Parma, Ohio 24153

Dear Sir:

Your job description for an executive secretary, placed in the November 1, 1975, *Cleveland Plain Dealer,* is very similar to the position I now have in Dearborn, Michigan. Since my wife has been promoted to an executive position with a large corporation in Cleveland, I am interested in working in that area. A–1 A–2 A–3

For the past six years, I have worked in the Teacher Education Department of the University of Michigan as an executive secretary for President Collins. My work experience includes shorthand, typing, several kinds of business machines, and the conducting of meetings related to a variety of university problems. These duties, along with different kinds of personnel problems, were delegated to me by the President. B–1 B–2

Before working at the University of Michigan, I was employed as a private secretary to a high school principal in a large Livonia, Michigan, high school. While there, I listened to and learned about problems youngsters have. I believe that my years of personal contact with high school students has given me a broad knowledge of their problems and what to do to help. B–3

Since I must move to the Cleveland area, I am especially interested in working in the Parma school system. I am sure my three elementary school children will be happy; they already have friends in Parma.

I would appreciate an interview at your convenience. I will be happy to send you my résumé or bring it with me at any time you designate. If there is additional information you need, please contact me at (313) 349-2455. C–1 C–2, 3, 4

Yours truly,

John O'Reilly

RÉSUMÉ OR DATA SHEET

The résumé or data sheet contains most of the information asked for on any employment application. Your prospective employers often may prefer to have you complete an application blank. If that is so, they will be glad to provide you with one you can complete and return with a cover letter emphasizing your qualifications.

Throughout your working years you should keep a "live," up-to-date, active résumé, on which you add changes in your employment and educational experiences shortly after they occur. The names of supervisors, dates of employment, some of your important duties and responsibilities, and sometimes the names of the organization you join should be added to it. Doing this will enable you to provide this information accurately later, whenever you may need it in applying for that important position you someday want. It will keep you from forgetting important details.

How is a résumé prepared? What does one contain? These questions are answered on the next few pages. A well-prepared résumé may be the key to your future; learn to prepare a good one.

Résumé Elements

What is a résumé? The word "résumé" comes from the French *résumé*, meaning "summary." Your résumé, therefore, will be a summary of your life history, especially your educational and work experiences. It will be a means by which you will advertise the qualifications you are offering for hire. In that sense, anyone's résumé is a sales brochure. Through it a person sells his or her skills and tries to achieve two things: (1) secure an interview and (2) be hired for the position or a related one.

The résumé should be prepared before you write the cover letter for it. Doing it then enables you to become aware of the features that make you competitive. The data sheet or résumé summarizes your qualifications, but the cover letter interprets the data in the résumé and focuses the reader's attention on those features that make you outstanding.

The résumé usually contains four main classifications of data:

1. Educational.
2. Employment.
3. Personal.
4. References.

Begin with the classification you want to emphasize. The references usually are placed last because they suggest something your prospective employer can do right away. They have telephone numbers and addresses, making it convenient for him or her to telephone or write the people listed for more information.

Because the person to whom you are writing knows that as a student you may not have extensive occupational experience, he or she may be more interested in your educational qualifications. Place your educational data first unless your employment experience is more important for a certain position. When your work experience is more valuable than your education in a certain field, place it first.

The person to whom you are applying will be more interested in either your educational or your work experience; therefore, place the personal data after them and the personal references last.

Résumé or data sheet

Here are some helpful tips on how you can make your résumé data more competitive:

1. Regardless what type of information you place first, make sure that your name, address, and telephone number are easy to locate somewhere in the heading.
2. Also in the heading, opposite the words "Employment Objective:" name the exact job or the type of job for which you are applying.
3. Place an appropriate photograph in the upper left- or right-hand corner of the data sheet. The photograph will reveal information about you that can be expressed in no other way. Consequently, the reader will feel he knows you better.
4. Under the Employment Data heading, identify each job, each employer, and explain the main duties for each. Emphasize the work responsibilities closely related to the "Employment Objective" of that particular data sheet.
5. Provide accurate information. Do not give deliberately misleading or exaggerated information. It may be better to remain silent about something instead of being dishonest. It is not unusual for an applicant to be rejected after the interview because his dishonesty is intuitively felt, suspected strongly, or clearly detected.
6. Do not hesitate to identify and give details about any special assignments or projects for which you were selected because of your knowledge or skills.
7. Modestly identify and briefly explain the important promotions received from employers.
8. Identify any special awards you have received from your employers or community organizations. Also, you might tell of your important hobbies and interests outside the job, to reveal your healthy, wholesome nature and attitude toward life. A happy employee is an asset; an unhappy one is a liability.
9. To give a good cross-sectional view of your experiences and character, submit a variety of references. For a character reference, select someone who knows you as a person, not only as a client, customer, or patient. For an academic reference, submit the name of an instructor from your major, especially one from a field related to the vacancy you want to fill. For testimony about your employment reliability and capabilities, offer the name of one or more of former employment supervisors. If you can, pick one whose skill in the job for which you are applying would be valued.
10. Place all the information on a single page, if you can, but remember that it is better to use more than one page to avoid giving the data sheet a crowded, hard-to-interpret quality.

Following is a format showing the details and organization usually required in preparing a résumé. Remember, whether the educational data or the employment data are placed first depends upon which have the most important qualifications for the type of employment objective shown in the heading of the résumé.

RÉSUMÉ (*or* Data Sheet)
December 20, 19__

Photograph

Name ________________
Address ________________
____
Telephone ________________

EMPLOYMENT OBJECTIVE: ________________
(Specifically identify the job or position—for example, truck mechanic, legal secretary, accountant)

SKILLS AND FIELDS OF COMPETENCE

Cite the titles of positions and the fields in which you have achieved degrees of skill and recognition for your competence—for example, state certification as an auto mechanic, investment consultant, chief chef in a well-known hotel, etc.

EDUCATIONAL DATA

(Show all levels in reverse chronological order by starting with the latest school attended and continuing back in time. "Employment Data" may be placed before this section if it is more advantageous to do so.)

1. Dates attended, name and address of each college or university.
2. Degrees received and honors granted, if any.
3. Major and number of credit hours completed (or include in a summary of areas of concentration).
4. Minor and number of credit hours completed.
5. Additional, especially related, courses or projects.
6. Relevant extracurricular hobbies and interests.
7. Any special awards and other forms of recognition for achievement (certificates, license).
8. Dates attended; name and addreess of each; technical, vocational, or business institutes or schools, giving a brief description of the courses and additional training undertaken.
9. Dates, branch of service, unit, and where special military training was acquired.
10. Dates attended, name and address for each high school, telling whether you took a college preparatory or a vocational course. There is no need to give junior or elementary education unless you just recently graduated from high school.

<u>Main Area of Concentration</u>: Under each graduate school, undergraduate college, university, or technical or vocational school, write a brief paragraph identifying your field of specialization and expand on special projects or extracurricular activities especially related to the duties and responsibilities you may be required to assume in the employment for which you are applying.

EMPLOYMENT DATA

(Place in reverse chronological order)

1. Dates, name, and address of each employer, including present one, if employed.
2. The title of each job and its duties and responsibilities.
3. Under each employer, provide a brief summary of your duties, emphasizing those relevant to the one for which you are applying.
4. Under each former employer, give the name and position of your immediate supervisor, if you can; at least give as many as you can remember.

Main Duties and Responsibilities: After each place of employment, as exactly as you can, cite the title of the position held and briefly discuss the duties you performed along with your responsibilities and the people to whom you were accountable. Emphasize with specific details the experience in each job that makes you especially qualified for the employment for which you are now applying.

PERSONAL DATA

1. Birth (date and place).
2. Physical condition (height, weight, general health, handicaps).
3. Marital status (single, married, number of dependents).
4. Military service (branch, dates, place of service).
5. Travel (willingness to relocate, willingness to travel, where, and to what extent?)

REFERENCES ("by permission," if granted)

Cite the names, addresses, and telephone numbers of at least three persons as references. Submit one who can support your character as a person; another, your ability and reliability as an employee; and one more, your educational achievement. Remember, it is courteous to ask a person for permission to use a name for reference. Only indicate that you have permission when you were able to get it, not when unable to get it. If you use your present employer as a reference, be sure to indicate that he is your present employer, and tell whether or not it would jeopardize your present job if he were contacted for information about you. You certainly don't want to lose your present job before you are sure of getting the one for which you are applying.

ORGANIZATIONAL MEMBERSHIP

List the names and addresses of the various professional, occupational, and important community organizations to which you now belong or have belonged. Be sure to indicate whether it is a current or a prior membership. If you were an officer or served on important committees, tell which positions you held and on which committees.

AWARDS AND FELLOWSHIP

Give the names of the awards, citations, scholarships, or fellowships you have received. Cite the dates you received them and the name of the person or organization who awarded them.

KNOWLEDGE OF LANGUAGES

If related to the position for which you are applying, identify the foreign languages with which you are familiar. Give your ability level in reading, writing, speaking, and understanding each.

It would be very worthwhile to indicate in the résumé cover letter or at the bottom of the résumé when you will be available to start work. If you are unemployed, tell the person to whom you are writing that you are able to start immediately. Also, indicate whether or not you are willing to travel on a regular basis should the job require it. Under either the occupational or the educational heading you may want to indicate military training or occupational experience to give your reader a more complete understanding of your background. Since the military draft is no longer in existence, there is no need to indicate your status.

The résumé does not have to be longer than one page, but do not leave out any worthwhile experience or education just to make it short.

Following is an example of a résumé written by a student whose occupational experience made him especially competitive for the position; consequently, he placed it before his educational training.

RÉSUMÉ
January 10, 19__

Photograph

William R. Rigsted
15007 Flamingo Avenue
Livonia, Michigan 48154
Telephone (313) 421-4495

EMPLOYMENT OBJECTIVE: Photographic Manager

SKILLS & FIELDS OF COMPETENCE

Supervisor of a motion picture and videotape unit. Illustrative photographer, producer of audiovisual presentations, production coordinator, photographic and videotape salesman, photographic lab experience.

EMPLOYMENT DATA

1973–present *Supervisor* of Videotape and Motion Picture Productions Dept., Ford Motor Company, Dearborn, Michigan.

Main Duties and Responsibilities: I am responsible for a 100% accountable unit that produces motion pictures and videotapes for management communications, employee training, merchandising, and advertising. My responsibilities include customer contact, scheduling, budgeting, production coordination, and maintenance of a million dollars worth of equipment.

1970–1973 *Creative Photographer*
Ford Motor Company, Dearborn, Michigan.

Main Duties and Responsibilities: As a creative photographer I worked on new car photography in the studio and on location, and on other illustrative assignments. For a year and a half I worked on the photography for the Ford Motor Fairlane Development and the Renaissance Center Development; this included 16mm films, five projector slide presentation, and photography for brochures.

1966–1970 *Industrial Photographer*
Ford Motor Company, Dearborn, Michigan

Main Duties and Responsibilities: As chief photographer in the Audio Visual Section, I assisted all levels of management in the planning and production of 35mm color slide presentations. This included client meetings, script interpretations, visual recommendations, illustrative studio and location color photography, and projection assists.

1963–1966 *Industrial Photographer*
Micromatic Home Corporation, Detroit, Michigan.

Main Duties and Responsibilities: I was responsible for the photographic department and worked with management on all projects. This would include new marketing ideas, planning and photography for sales seminars and technical presentations, industrial movies, advertising catalog illustrations, research photography and related services.

1959–1963 *Photographic Illustrator*
Creative Services, Inc., Detroit, Michigan.

Main Duties and Responsibilities: I photographed slide films in color and black and white, produced slides for advertising agencies, worked on the photography for agency ads, and assisted in the photography of record album covers.

1956–1959 *Photographer and Lab Supervisor*
United States Army—Korea.

Main Duties and Responsibilities: My assignments included service pictures, feature, sports and publicity photographs for the Corps newspaper and aerial photographs. I also supervised four other photographers in the Corps photo lab.

EDUCATIONAL DATA

1969–present Schoolcraft College, Livonia, Michigan.

Areas of Concentration: Two years of college credit toward a degree in Marketing. All classes have been taken at night while working full time in my present position involving photography. Additional classes were also taken to gain knowledge and additional skills in commercial photography.

1956–1957 U.S. Army, Forth Monmouth, New Jersey.

Areas of Concentration: I attended still photographic school at Fort Monmouth, New Jersey.

1952–1956 Cooley High School, Detroit, Michigan.

Areas of Concentration: After taking a college preparatory course I was graduated in 1956. During my high school years, I worked on the school newspaper taking photographs.

PERSONAL DATA

Age:	36
Marital Status:	Married, 3 children
Health:	Excellent
Height:	6′ 0″
Weight:	180
Relocation:	Yes, anywhere in the U.S. only.
Military Service:	Three years with the U.S. Army; honorable discharge in 1959.

PROFESSIONAL MEMBERSHIP

Industrial Photographers Association of Michigan—Board member for three years.

Academy of Video Communications—Member of Committee on Meetings and Conventions.

AWARDS

Numerous letters of commendation for projects on which I have worked.

REFERENCES (By permission)

Mr. McHenry (Employment)
Ford Motor Company
Dearborn, Michigan 44168
Telephone No. (313) 622-7419

Professor Edward Roray (Academic)
Michigan State University
Lansing, Michigan, 40414
Telephone No. 471 (414) 472-9040

Ms. Mary Langston (Character)
1582 Kingsmill Road
Bloomfield, Michigan 45194
Telephone No. (313) 619-1823

Résumé Cover Letter

A résumé cover letter performs somewhat the same functions as the employment application letter. The main difference is that the résumé cover letter introduces a résumé, a data sheet, giving the main details about the applicant's educational background and work experience. The cover letter itself, therefore, need not have much of this information. It mainly focuses the potential employer's attention on the applicant's qualifications especially related to this particular job.

Résumé or data sheet

Always enclose a cover letter when mailing out a résumé. Your main purpose is to arouse your possible employer's interest in hiring you or calling you for an interview. Your first concern, therefore, is to get him or her to read your résumé. Keep these thoughts in mind when writing the cover letter for a résumé. Here are some tips to help you write a good one:

1. Address your letter to a specific person by name, when possible.
2. The first twenty words are important; they should attract the reader's attention.
3. Tell your story in terms of the contribution you can make to the employer.
4. Be sure to refer to your résumé. It gives the facts.
5. With local firms, take the initiative in suggesting that you may be scheduled for an interview by telephone.
6. Use simple, direct language, and correct grammar. Avoid overused expressions, and, of course, type neatly.
7. Keep it short; you need not cover the same ground as your résumé. Your letter should sum up what you have to offer and act as an "introduction card" for your résumé.
8. Let your letter reflect your individuality, but avoid appearing aggressive, overbearing, familiar, "cute," or "humorous." You are writing to a stranger about a subject that is serious to both of you.
9. Point out important conclusions and qualifications in the résumé that are related to the position for which you are applying.
10. Insert into the letter any experiences not mentioned in the résumé that are related to the position.

Below is a suggested outline you may use as a guide when writing a résumé cover letter. Notice the suggestion in the "B" Section that your strong qualifications related to this particular opening be emphasized.

A. *Beginning Section*

1. State how you learned about the opening: newspaper ad, employment agency, a friend or relative.
2. Tell why you are interested in working for that particular company.
3. State that you are enclosing a résumé.

B. *Developing and Supporting Section*

1. Clearly point out the qualifications on the résumé that make you a good choice.
2. Focus additional attention on any special projects you completed, awards received, work experience, and any experience in operating a business of your own.

C. *Concluding Section*

1. Request an interview at the receiver's convenience.
2. Clearly tell where you may be reached, including your telephone number.
3. Tell whether or not he or she may contact your present employer for reference.

The résumé cover letter below was written to accompany the foregoing résumé.

15007 Flamingo
Livonia, Michigan 48154
December 14, 1974

Mr. D. L. Russell
Employment Manager
Mark Motor Company
American Road
Dearborn, Michigan 48212

Dear Mr. Russell:

I read with interest your ad for a photographic manager in the Sunday, December 12, 1974, *Detroit News*. Therefore, I am submitting my résumé for your consideration. A–1 A–3

Your Photomedia Department has a worldwide reputation for its progressiveness, and I have always wanted to work in such an imaginative organization. Please notice on the enclosed résumé the extensive experience I have had in supervising motion picture, videotape, and illustrative photography. A–2 B–1

Presently I am supervising a motion picture and videotape production unit of twelve men. My strongest qualification is motivating employees toward group objectives rather than only individual achievements. I have experience in producing management communications, employee training, merchandising, and advertising productions. These abilities are not easily detected on the résumé I am sending you with this letter. I also want to point out something else not included in it, I ran my own illustrative photography studio in Chicago for three years. B–2

I feel I am well qualified for this position, and I am eager to discuss it with you further. I would appreciate an interview at your convenience. You may contact me at my home address, or telephone 421-6395. I would prefer you not call my present employer in reference to your position. C–1 C–2 C–3

Yours truly,

William Rigsted

WR:jk
Enc.

THE INTERVIEW

The main purpose of your résumé is to persuade the prospective employer to grant you an interview. The résumé itself later serves as the basis for the interview. It prepares you for it by enabling you to anticipate some questions that may be asked. Also, it provides the interviewer with some interesting areas to be explored.

The interview

If your cover letter and résumé succeed in persuading someone to call you in for an interview, you will have achieved an important step in getting the position you want. In comparison to the number of people applying for a good position, only a few of them are given an interview. If you are one of these few, your letter and résumé will have done a good job. Whether or not you get the position will depend then on the outcome of your interview.

An interview for employment is not an interrogation, a cross-examination, or the third-degree. It is not intended to put you on the defensive. The person conducting the interview is mainly aiming to get enough information about you to enable him or her to make an intelligent decision. You, therefore, should participate in it cooperatively.

As soon as you can after sending your résumé, you should start preparing yourself for the anticipated interview. Get the right attitude toward the company and its products or services by doing a little research. Here are some questions you may be expected to answer for yourself before arriving for the interview. If you are hesitant or uncertain about them, your interviewer may wonder whether or not you honestly want the job, and how long you will stay even if you do get it.

1. Will my experience and training enable me to do the work as well as I am expected?
2. Are there promotion opportunities related to my ambitions?
3. How much pressure will I be under, and can I work well under that much pressure?
4. Is my attitude toward the company and its products wholesome?
5. Are the working conditions safe, pleasant, and encouraging?
6. Will I have to relocate, and will those who depend on me be happy with the changes?

During this interview you try to prove you are all that you say you are in your résumé. Your interviewer is eager to have you do so; he or she will even ask you leading questions to help you reveal your potential value. One main purpose of the interview is to see if your personality and qualifications will enable you to mesh with the nature and objectives of the organization thinking about hiring you. Along with asking questions about the information in your résumé, the interviewer will observe your behavioral responses. The amount of confidence you have, how relaxed your speech and movements are, your friendly or belligerent attitude, and how clearly you express yourself will influence the conclusions about you.

Here are a few important tips to guide you in presenting yourself best during the interview:

1. Be businesslike and professional in appearance.
2. Don't show overanxiousness.
3. Don't sprawl in your seat. Just sit naturally. Being stiff will reflect nervousness and a lack of confidence.
4. Do not smoke unless given permission.

5. Be courteous. Do not interrupt your interviewer.
6. Don't talk too much. Let the interviewer ask the questions for which he or she wants answers.
7. Listen well with your eyes and ears.
8. Show sincere, real interest in the welfare of the company.
9. Be very direct in your statements; if you do not know the answer, say so.
10. Show self-confidence, but not cockiness.

Some wrong impressions about you may result from your being nervous or from trying to cover up your nervousness. Expect to be nervous and uncomfortable at the beginning of the interview; that's natural. Experienced interviewers try to help you regain confidence by creating a friendly relationship at the beginning so you will be able to give the needed information. Overcome your queasy feelings by keeping your mind focused on the questions asked and responding to them effectively. Doing that kind of concentration will help create an understanding that you and the interviewer are on the same side. You will begin to know that you are not working against each other, one probing and the other defending, and your nervousness about having weak spots discovered will vanish.

Don't be dishonest, evasive, or intentionally unclear in answering questions. Be especially careful not to blame others for the mistakes you made or for your inability to achieve promotions in your other places of employment. Occasionally an interviewer may ask startling questions to see how well a person behaves under pressure when applying for a job that requires stability—perhaps a question like, "How do you feel about abortion?" This question could have various answers. If you haven't thought about it much and don't have an honest opinion, say so. If you don't understand the question as exactly as you should, ask for explanations. Ask questions such as, "Do you mean for married or unmarried people, for young newly married or for much older people, for sick or handicapped persons or those with excellent health?" You might further ask, "Do you mean medically, morally, or ethically?" The important thing is to show the interviewer that you are willing to respond once you know what is wanted, and that you calmly keep your eye on your objectives even under severe pressure.

Following are some of the criticisms most often made by personnel people who conduct many interviews for employment. They say the people they interview often:

1. Can't express themselves calmly, accurately, and clearly.
2. Have belligerent, defiant, defensive attitudes.
3. Are indecisive, uncertain, or lack confidence.
4. Are mainly concerned about wages and are shopping around for jobs requiring the least effort.
5. Are not honest about their qualifications.
6. Are overly ambitious, not realistic about their abilities.

The interview

Because good interviewers are eager to match the right people with the right jobs, they are happy to answer questions about a position for which you are applying. The questions may tell them certain things they want to know about you, about the kind of person you are: your values, ambitions, interests, and how well you get along with people. Therefore, ask your questions sincerely and thoughtfully. Ask questions that will help you get the information you need to make an intelligent decision about whether or not you really want the job, now that you understand better what it is.

Here are some questions you may want to put to your interviewer:

1. Judging from my résumé, do you think I am qualified to do the work well enough?
2. Will I receive any training or education to help me strengthen my qualifications?
3. How do you evaluate employees for promotion?
4. How often and what kinds of wage increases should I expect?
5. What kinds of fringe benefits, other than money, will I receive?
6. What are the possibilities that I will be expected to move to another city and where?

It is not unusual to be told during the first interview whether or not you have been selected for a position. More often, however, the job applicant is told that the résumé and results of the interview will be given to another person for a decision. This is especially true if the person for whom you will work is someone other than the interviewer. In either case, however, your success will depend upon the skills with which you prepared your résumé and cover letter, and on how well you participated in the interview.

Application 11–1

Following are several letters of application for employment. In the margin to the right of each, identify the main sections each contains by inserting the correct letter (A, B, or C). Then insert the subsection numbers. These letters and numbers should be placed just as they are shown in the discussion illustration in this chapter.

A. Dear Sir:

I read the position description for systems engineers in the 1974 College Placement manual and thought it was very interesting and quite challenging. I would like to secure this position for several reasons: the position deals in an area which I find very interesting and correlates quite well with my educational and professional background. I spent ten weeks in Treasure Island during my tour in the Navy, and I would very much like to settle down in the San Francisco Bay area. I would like to expand upon the experience and tenure I had acquired while serving in the U.S. Navy.

During my 41-month tour of active duty, I had received a considerable amount of training and experience in various areas pertaining to the shipboard engineering plant, systems, and operational procedures. As the Engineering Officer onboard a Naval Combatant, I was responsible for the efficient management and operation of the men and material in the Engineering Department. With this experience, I developed insight into the methods of shipboard engineers and the problems encountered during operations at sea. I feel this experience would effectively enhance my ability to analyze problems and to design systems for use onboard naval vessels.

Should my application merit further consideration, I will be at your disposal and available for an interview any time during the months of January and February. If you require additional information, I can provide my service fitness reports and my academic transcripts. I can be reached at the address and number listed on the résumé. Your attention to this application will be very much appreciated, and I will be anxiously awaiting your reply.

Respectfully,

B. Dear Sir:

I am writing you to inquire about the possibility of future employment at the X-Motors Corporation Proving Grounds complex, Milford, Michigan. I would like to inquire specifically about a position as a automotive technician in the Department of Research and Development.

At the present time I am working at the Y-Motor Company Proving Ground, Romeo, Michigan. My past working experience includes three years as an automotive technician at Jam Bivan Agency in Plymouth, Michigan.

My educational background consists of a high school diploma from Henry Ford School in Westland, Michigan, an Associate Degree in Automotive Service Management from ZZ Community College, a Bachelor of Arts Degree in Automotive Technology from Ferris State College.

I have heard a great deal about the different phases of research and development being done at the X-Motors Proving Ground, and would like very much to have the opportunity to be employed by you as an automotive technician anywhere.

I would be very grateful for any information that you can send me concerning the X-Motors Proving Ground, and I will be eagerly awaiting your reply about possible future employment. I will be glad to send any other information you need.

Sincerely yours,

C. Dear Mr. Burton:

Your senior electrical engineer, Thomas Mills, has informed me that your firm has an opening for a mechanical engineer specializing in pollution control. I would be very happy to work for your company because of the excellent working conditions and engineering freedom there.

As the enclosed data sheet indicates, I am well qualified for this pollution control position. Not shown on this résumé, however, is experience in pollution systems testing. While at Bolin State University, I assisted Dr. Leonard Quincy in his research with his well-known ABO system. I feel this experience makes me especially qualified for this mechanical engineer position.

I would very much like a personal interview at your earliest convenience. I may be reached at 721-4611 from 8 a.m. to 5 p.m. on weekdays and at 349-6712 other times. I would prefer you didn't contact my present employer as it might jeopardize my present commitment. If you need any additional information, please feel free to contact me. I am very confident I am qualified for this position and will be eagerly awaiting your reply.

Sincerely,

Application 11–2 Do the same for the two following résumé cover letters as you did for the letters in the preceding application.

A. Dear Sir:

I learned of your need for a Division Manager from Mr. L. M. Harkens. I am very interested in working for your organization since it is a leader in the industry. The position I would like is that of Division Manager; however, if that position is already filled, I would consider one leading to it.

I would like to point out my many years of experience in the various facets of furnishing prefabricated piping. In addition, I feel my name and reputation are fairly well known with many of your customers which we at Utica have supplied. One of the last projects I was directly involved with at Utica Engineering, as Manager of Engineering, was the implementation of a production control system with the use of a computer. The system was instituted at the bid stage, utilized to print master bills of material, maintain inventory control, labor load at various cost centers such as cutting, machining, bending, fabrication, welding, NDE, cleaning, printing, and shipping, billing after shipment, and keeping up with costs per labor operation, as well as profit and loss by the piece and by the job.

I would like to arrange an interview with you at your convenience to discuss my qualifications for this position. During the week, please call my home 321-4231 to arrange an interview or to request any additional information which may have been overlooked. I would appreciate it if you would consider this letter confidential. I feel confident that I am qualified for the position and am eagerly awaiting your reply.

Very truly yours,

B. Dear Mr. Robie:

As you suggested in our meeting at the Conference of Metal Foremen in Chicago, Illinois, on February 3 ,1974, I am submitting the enclosed résumé, supporting my application for employment as an alloy foreman in your Rapid City, Iowa, plant.

Our conversation at the Conference allows me to feel that my extensive experience in supervising the production of brass and copper alloys will enable me to be especially helpful in making your products more competitive.

I will be able to travel to Rapid City whenever an interview with you will be most convenient. If you wish, you may telephone me at 477-8121.

Yours truly,

Application 11–3

Write a letter of application for a job you would like to have. Now, write a favorable reply to the letter of application, asking for a résumé and agreement to come for an interview later.

In the margin at the right of each letter, insert the capital letters and the numbers that identify each main section and subsection. Use the illustration letters in this chapter as a guide.

Also, opposite the letter on a line at the bottom, insert the capital letters and numbers of the subsections intentionally not included in it.

Application 11–4

Write a résumé and a résumé cover letter for the letter requesting them in the preceding application. In the margins, identify the main sections and the subsections.

Application 11–5

Get an application blank from a firm for which you might like to work, read the questions on it, and make a list of questions that you might ask about the information it supplies.

Next, make a list of questions you think you might be asked during a subsequent interview to explain or support your application answers.

Application 11–6

Contact a person who conducts employment interviews in the field in which you are or hope to be employed. Ask what he or she looks for in an interview and what techniques are used in getting the needed information. Write a short paper identifying what you found out and explain whether or not you think the interviewer can make a good choice based on these techniques and the resulting information.

12

Research and organization

Informative vs. interpretive reports . . . determining your specific topic . . . brainstorming for ideas . . . using your library . . . finding reliable information . . . gathering and using notes . . . the parts of a report . . . the body of a technical report . . . bibliography and footnotes

This book tells and shows you how to write the kinds of reports you may be expected to read and to write as a part of your job. Although long formal reports are important in business and industry, seasoned technical report writers usually are assigned to write them. Long formal technical reports often are printed, expensive, and require several people and considerable time to prepare. Consequently, they are often written by outside agencies specializing in their preparation.

In this book we are mainly concerned with the shorter technical report. This is the kind of report most people are expected to know something about. People who are mainly concerned with important duties other than writing—engineers, contractors, technicians of all kinds—are often required to write some kind of short technical report. Anyone in a supervisory position may be expected to write one related to his or her main duties.

While instructing you in the writing of the shorter technical report, this book will also be showing you how to write the important sections of the long report. By writing and putting several of these sections together, should you ever be required to do so, you will be able to prepare an effective long report.

INFORMATIVE VS. INTERPRETIVE REPORTS

Dividing all technical reports into two general types will make your study of them easier. You can sort all technical reports into the two following classes, depending upon the kind of response you want from your reader:

1. Informative—To provide readers with new or different reliable information from qualified sources.
2. Interpretive—To present conclusions, including value judgments and recommended problem solutions.

The main purpose of informative reports is to give the intended readers accurate and reliable information. Your readers may accept this information just to get a better understanding of a technical subject, or they may interpret the information to arrive at their own conclusions and judgments. Based on what your report tells them, they may make important technical decisions. Therefore, you should constantly check the accuracy of your statements along with the reliability and qualifications of your sources.

In an informative report you should not include your own interpretations, conclusions, evaluations, or recommendations. When the person who assigns the report wants interpretations from you, he or she will want you to write a different kind of report, an interpretive one. It is important to know the difference, so that you will write the kind of report wanted, whether or not it is specifically assigned.

Interpretive reports should present more than the reliable information you gathered from your own experience and from researching what other qualified sources have written. In addition they should include your evaluation of the information you collected and explain the conclusions that follow from your thinking. These conclusions may be your value judgments about the merits of something, such as a piece of equipment or a way of doing something; they may be your solutions for a problem; or they may be just the results of your investigation, reporting what you found to be the causes for a problem.

DETERMINING YOUR SPECIFIC TOPIC AND MAIN IDEA

You can do very little technical writing without knowing exactly what you are going to write about. Even to do much preliminary thinking or research you need to know your specific topic and main idea. Sometimes it is a simple matter. If, for example, you are called into a superior's office and asked to investigate and explain in a report why a certain hot water pump is not working well, all you have to do is to start

Determining your specific topic and main idea

making your own observations. Usually you don't have to search for what others have said about that specific problem. You know what your specific topic is before you start. You also know that the main idea will be whatever your findings are, the cause or causes of the problem. For example, your specific topic and main idea might be: "The hot water pump for the radiator tester has a burned-out electric coil; consequently, it can't provide the needed quantity of high-temperature hot water." For this main idea, all you will have to do is write an informative report giving the status of the pump. The person reading it will interpret your information and arrive at his or her own conclusions and decisions.

However, if your supervisor asked you not only to find out why the pump wasn't working well but also to make recommendations for solving the problem, you would have to write an interpretive report, and your main idea about the pump would be different. Perhaps it would be something like the following: "Because the electric coils in the hot water tank we are now using cannot withstand the high temperatures for extensive periods, the water for the radiator tester should be supplied by a tank with a gas burner." From this, it is easy to see that the main idea will help you determine the kind of report you have to write, when you are not specifically told which kind. Sometimes it may be necessary to ask which kind is wanted.

When you are told, as in the preceding illustration about the pump, what your specific topic is, writing a report is made easier. All you have to do next is start your first-hand investigation. More than likely, you won't have to go to any books; you can get most of what you need by making your own close observations. While you are in college, however, you are not assigned that specific a topic. You usually are given a much broader one and told to go from there, and the hard part of selecting a specific topic and arriving at a main idea about it is left up to you.

BRAINSTORMING FOR IDEAS

Brainstorming is one effective way of selecting a worthwhile specific topic and main idea. This is a process of free association in which you introduce an idea into your mind and write down the ideas it calls up. Brainstorming was applied extensively during World War II in solving some difficult problems, and it is still being used.

Students can brainstorm worthwhile technical ideas for technical report writing. When you already have an extensive knowledge of your area, you can just sit down with a pencil and paper and jot down subtopics and main ideas related to the first one as they occur to you. When you think you have enough, you can go over your list and evaluate them to select a good one for the kind of report you intend.

One good way of arousing free association in a general subject area is by leafing through the card catalog in your library. Read the titles and, perhaps, the notes that describe each book's contents. As possible ideas for your report occur to you, jot them down. You can do the same thing with the lists of books at the end of the main articles in various encyclopedias. Another good place to brainstorm worthwhile ideas is in *Reader's Guide to Periodical Literature* or the several indexes to technical journals.

Let's suppose your geology instructor assigned the general topic "rocks" and asked you to write an informative report of some kind. In trying to brainstorm a specific topic and main idea, you found the following titles listed in the sources indicated:

Reader's Guide to Periodical Literature (March 1973 to February 1975)

Have Some Fun with Garden Rocks
Carbonated Rocks

Library Card Catalog

Geology of Industrial Rocks and Minerals
How to Know the Minerals in Rocks
Sequence of Layered Rocks
The Evolution of Igneous Rocks

Encyclopedia Americana, Vol. 23

Deposition of Sedimentary Rocks
Basaltic Rock of Volcanoes
Architecture of the Earth
Rocks and Minerals
Shale Oil

Just by letting your mind wander over the above titles you might get several ideas to jot down, such as the following:

1. Role of volcanoes in the architecture of the earth.
2. How to identify worthwhile industrial minerals in rocks.
3. How objects found in rocks can be dated by the sequence of rock layers.
4. Proposition: our total industrial needs for crude oil can be supplied from shale without serious surface damage.

Now that you have a series of specific topics and main ideas from which to choose, you can select one you think your reader will consider worthwhile. The main idea you select will suggest the kind of reader response you will aim at and the kind of report that best will enable you to achieve it. The first three ideas in the above list require just informing the reader; therefore, you will want to write any of several kinds of informative reports. The last one, however, will require your persuading the reader to accept your conclusions; therefore, you will have to write an interpretive report containing your interpretations, conclusions, and recommendations for solution of an important problem.

USING YOUR LIBRARY

Libraries are the main source for information about technical ideas. Whenever you can't make your own first-hand observations to get the information you need, you will have to search in a book, article, or some other kind of communication for reports of direct observations made by others. Students, of course, have neither the time nor the money needed to conduct the extensive studies often required to get the information they need. But why should any researcher, even in the biggest industry, spend time, cash, materials, or labor in making an investigation already completed and reported on by someone else? For this reason, libraries have become an important industrial as well as personal resource.

The Dewey Decimal classification (a system of numbers) and the Library of Congress system (numbers and letters) are the two ways libraries classify and shelve books. Most libraries use the Dewey classification, which is suitable for book collections both large and small. Large libraries, such as those in very large universities, use the Library of Congress system.

The Dewey system gathers books on the same subject side by side. Its main classifications are as follows:

000	General Works	500	Science
100	Philosophy, Psychology	600	Applied Science and Technology
200	Religion	700	The Arts
300	Social Sciences	800	Literature
400	Language	900	History, Geography

An "R" above the number indicates the book is in the reference section of a library. Reference books are generally authoritative sources of reliable information; they do not circulate. After you have used the reference books related to your occupation, you will come to recognize which are dependable.

Among the most often used reference books are the indexes to periodicals. It would be almost impossible to find a certain article about a technical subject published in one of the many newspapers, magazines, and technical journals without the help of an index. *The Reader's Guide to Periodical Literature* is the most familiar; it lists the contents of around 125 popular magazines in much the same way the more technical indexes do.

Library Card Catalog

Usually, each nonfiction book is represented by at least three cards in the card catalog: one for the author, one for the title, and one or more for the subject.

Much information is given on a catalog card, as shown by the author card illustrated here.

AUTHOR CARD

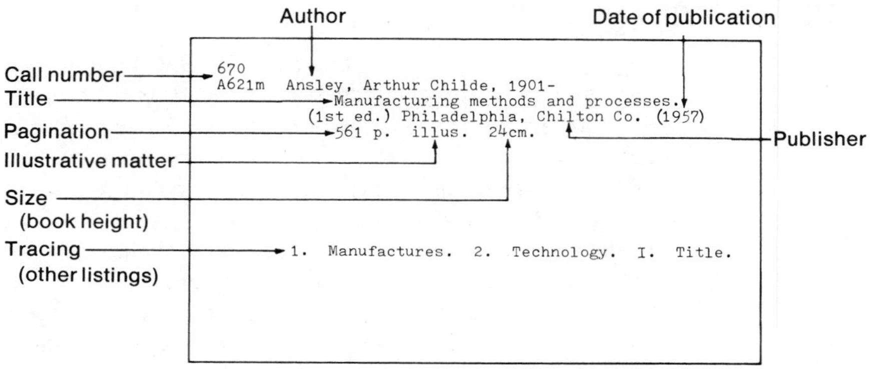

The title card looks exactly like the author card, except that the title is inserted above the author's name. The card is filed under the first word of the title, not under an article: *a, an, the*, etc.

TITLE CARD

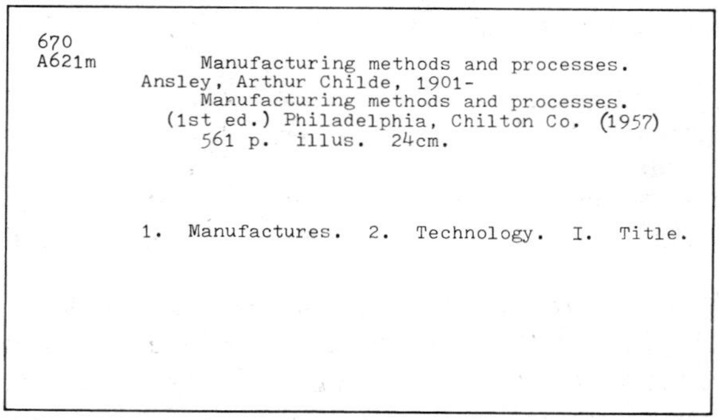
670
A621m Manufacturing methods and processes.
Ansley, Arthur Childe, 1901-
Manufacturing methods and processes.
(1st ed.) Philadelphia, Chilton Co. (1957)
561 p. illus. 24cm.
1. Manufactures. 2. Technology. I. Title.

When you do not know the exact author or title, you may find a book filed on one or more subject cards. On these cards the subject of the book is typed in red above the author's name. Some libraries use capital letters instead of red ink. You should go from the specific subject to the general when using the subject catalog, not the other way around. When the subject is elephants, first look under elephants and then under animals of India or Africa.

As many different subject cards are in the card catalog file as are needed to cover the material in a book. Illustrated here is a subject card for the book by Arthur Childe Ansley. Notice the subject heading above the author's name.

SUBJECT CARD

670
A621m Manufactures
Ansley, Arthur Childe, 1901-
Manufacturing methods and processes.
(1st ed.) Philadelphia, Chilton Co. (1957)
561pp. illus. 24cm.

1. Manufactures. 2. Technology. I. Title.

Using the Indexes

Each listing of the *Reader's Guide* has the following information:

1. Title of article.
2. Author.
3. Description of the article: illustrations, maps, etc.
4. Title of the magazine (abbreviated).
5. Volume number.
6. Pages on which the article appears.
7. Date of the issue.

Following is a typical subject entry:

Lebanon

Phoenicia, Abdo Haddad. il Holiday 23: 110-32, Ja. '75.

In the author entry, it is the author's name that is printed on a line by itself:

Haddad, Abdo

Phoenicia, il Holiday 23: 110-32 Ja. '75.

The preceding entries tell that the article entitled "Phoenicia" by Abdo Haddad is about Lebanon and was published in *Holiday Magazine* in Volume 23 in January 1975. These entries also indicate that the article contains illustrations (probably photographs) and was printed inclusively on pages 110–32.

Following are some indexes especially important to the technical writer:

1. *The Social Sciences and Humanities Index.*
2. *Biological and Agricultural Science Index.*
3. *Art Index* (includes articles on sculpture, ceramics, graphic arts, architecture, archaeology, and landscape architecture).

4. *Industrial Arts Index* (concerned with sciences in industry, engineering, trade, and business periodicals).
5. *Business Periodical Index* (indexes about 126 periodicals dealing with advertising, accounting, banking and finance, marketing, labor and management, general business, and insurance).
6. *Applied Science and Technology Index* (lists articles from about two hundred publications about aeronautics, automation, astronomy, chemistry, physics, construction engineering, geology, and related subjects).
7. *Engineering Index* (reviews about a thousand journals and has an abstract or summary for each article listed).
8. *Cumulative Index to Nursing Literature* (gives separate subject and author index to important nursing articles).
9. *Abridged Index Medicus* (separately lists subject and author index to major articles about medicine).
10. *Cumulative Index to Hospital Literature* (contains separate subject and author listings of articles about hospital operation).
11. *New York Times Index.*

Like the other indexes, the *New York Times Index* consists of headings and their subdivisions arranged alphabetically; the entries under these headings are arranged chronologically. Each entry gives the date, page, and column of the newspaper issue in which the story appeared. Each issue of the newspaper may be on microfilm in your library.

The *New York Times Index* has one important feature that makes it different from the others. Along with the other details about an article, it also has abstracts of many of them. A technical writer should know what abstracts are and how to use them as labor-saving devices. They are either brief descriptions, or else brief statements summarizing the important points, of an article or report. Here are examples of both kinds:

Descriptive Abstract

BIRTH Control and Planned Parenthood. Note: Material here deals largely with the medical, legal, moral and social aspects as they concern the individual and the individual family. For the link between birth control and the 'population explosion,' and for government efforts to curb the population growth by distributing birth control devices and information, **see** Population and Vital Statistics. Specific geog headings

Summary Abstract

BLOOD Pressure. See also Heart (for gen material on cardiovascular ailments)

Dr. Herbert Benson, cardiologist at Beth Israel Hosp, Boston, devises simple method of inducing relaxation and its attendant potential benefits to physical and psychological health; says his studies indicate that, if practiced regularly, relaxation exercise can produce significant lowering of blood pressure and improve a person's sense of well-being and ability to cope with world around him; says goal of his efforts is to find widely applicable method of counteracting adverse health effects produced in most people by continued reactions to environmental stress; describes classic response to stress as 'fight or flight,' in which heart rate, blood pressure and body metabolism increase and blood flows preferentially to the muscles; describes exercises (M), N 8,31:1

Often it is expensive in time and money to make lengthy on-the-job investigations to solve a complex problem or to write your own solution for one. Many times another person may have tangled with the same problem and published a good solution for it. But you don't have the time to read all the reports in your field to find that solution. However, by using technical abstracts, you will be able to examine many reports to see which have what you want. Anyone who must keep abreast of new technical developments, as well as students needing reliable information for research reports and other kinds of technical writing, must learn to use a variety of abstracts on the reference shelves of a library. You may find the following abstract indexes helpful:

1. *Chemical Abstracts.*
2. *Science Abstracts* (Section A is devoted to physics and Section B to electrical engineering).
3. *Nuclear Science Abstracts.*
4. *Biological Abstracts* (includes medical science, plants, animals, and related subjects).
5. *Historical Abstracts* (contains summaries of articles on political, social, and economic events worldwide).
6. *Abstracts of North American Geology.*
7. *Psychological Abstracts.*

In addition to the various indexes and abstracts, the library offers the technical writer a large assortment of dictionaries, directories, and encyclopedias. Become acquainted with them and learn to use as many as you find helpful. Here are some subject encyclopedias that may be very helpful to the technical communicator:

Encyclopaedic Dictionary of Physics
Encyclopedia of Mental Health
Grizimek's Animal Life Encyclopedia
McGraw-Hill Encyclopedia of Science and Technology

Microfilms and Microfiche

Two other important sources for a large amount of technical information are in your library. When you can't find what you need on the reference shelves, ask the reference librarian for the microform materials. There are two main kinds: *microfilms* and *microfiche.*

Vast amounts of reference periodicals are now reduced in size and photographed on reels of film. These microfilm reels are stored usually near the reference area in a library, close to the microfilm readers. Readers are machines that enlarge the film and light it from the back, enabling you to locate and focus the specific article you want.

Microfiche are cards, not reels, on which the print is reduced to 1/19 of the printed size, allowing many pages to be condensed to fit on a 4 × 6 inch card. These cards are also made of transparent film through which light is passed into an enlarging device called a microfiche reader. Both microfilm and microfiche have nearby indexes to help you locate the exact reels or cards you may need.

FINDING RELIABLE INFORMATION

After you have determined what you are going to write your report about and the way you are going to write it, including the kind of report, you will be ready to start searching for reliable information. A researcher can find a large amount of printed material about almost any topic. Because something is in print doesn't automatically make it reliable. In judging the value and reliability of ideas, you should always check the qualifications of sources. Here are some things to consider in deciding the qualifications of a source and the reliability of the ideas expressed:

1. *Date of publication.* Today's rapid changes in every field may make something true today false or not exactly true tomorrow. The copyright date, therefore, is an important criterion in judging the reliability of a source.
2. *Documentation.* Observing the quality of the sources listed in the bibliography and in the footnotes will indicate clearly whether the writer used qualified and reliable sources for the material. If he or she used second-rate sources, then that source must be researched and used cautiously.
3. *Publisher.* As in any other business, some publishers are reputable and some disreputable. The mere appearance of information in print doesn't mean it is reliable nor that it was written by a qualified person. A research writer must learn to distinguish between first- and second-rate publishers as an indication of the value of the ideas found in their publications.
4. *Qualifications of the author.* There are both recognized authorities and quacks among authors. By checking the experience background and the academic degrees of the author of a book or magazine article, usually the writer of a research paper or technical report can accurately judge the reliability of a source's ideas.
5. *Table of Contents.* Quickly examining the chapter headings or the table of contents of a book will indicate whether or not the ideas related to the thesis of the research paper are adequately treated to enable the objectives to be achieved.

GATHERING AND USING NOTES

The way you gather information while doing your library research is an important part of report writing. Collecting your findings efficiently on note cards will help you in organizing and writing your report later.

The first thing you should do when examining material related to your main idea is to write a bibliography card to identify your source. The information on this card will enable you to do two important things: (1) provide the information needed for your bibliography, identifying where you got the information for the whole report, and (2) provide the information needed for your footnotes. These bibliography cards, therefore, should be prepared carefully.

Take your notes on the same size card as you use for bibliography cards. This will make it easier for you to handle and arrange the cards when using them to write the first draft of your report. Your stack of bibliography cards should be kept separate from your note cards.

When you find reliable information that you think may be useful later, first head your card properly to tell exactly where you found it. This is essential, since you may have to go back to check accuracy, to get more information about a point you are making, or to identify the source in a footnote. The heading on each note card should include the last name of the author and the number of the page on which you found the note. If your bibliography contains more than one book by the same author or more than one author with the same last name, add a main word from the title to the last name. The last name of the author will act as a cross-reference to the bibliography card having the complete author's name, the title, and the publishing information. Should you need any of this information while writing the first draft, you can easily refer to it on the bibliography card.

The note on each card should be no larger than what you will need for use in a single place in your report. Don't take a note that has to be fragmented and used in several pages. Avoid taking bulky notes that deal with more than one important point you want to make clear later in the paper.

Now that you have your note card headed, you can decide which of the following kinds of notes to take: (1) quotation for accuracy or to demonstrate author's style, (2) summary of a series of related ideas or a chain of logically interdependent ideas, (3) a single idea in the note taker's own words.

Illustrated here is the bibliography card for notes shown on the cards printed on pages 201 and 202. Later, you will notice how the headings on the note cards refer back to the bibliography cards.

BIBLIOGRAPHY CARD

Author →	Thomas, W. C.
Title →	Nursing Homes and Public Policy
Publisher →	New York: Cornell University Press, 1969

Below is an extract from a report on nursing home regulation. Carefully observe the different kinds of notes on the cards following this extract and how each was used in writing parts of the report.

See page 5 note card

When the federal government moved into aid for the aged in 1935 with the passage of the Social Security Act, the social significance of a higher level of care for the elderly became recognized as an important concern of both state and federal governments. When federal funds became available to the states, amounting to half of what the states spent for persons 65 years of age or over, many people being cared for by their families sought admission into public and private nursing homes.

Resulting from this rapid increase in the number of aged seeking aid and with the shortage of hospital beds brought about by World War II, private nursing homes multiplied. This expansion began with private family homes but soon turned into proprietary (privately owned, profit-making) nursing homes. These proprietary homes increased in number dramatically when a law was passed in 1944 reimbursing states for payments made to these private nursing homes.

See page 12 note card

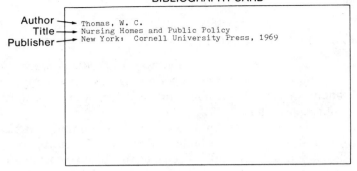

Care for the elderly passed from the nineteenth century almshouse, poorhouse stage to a more humane care with the enactment of legislation requiring registration, inspection, and licensing of both private and public nursing homes. This began when the Hill-Burton Hospital Survey and Construction Act of 1946 provided funds to assist in the building of these nursing homes. However, it only provided funds to build public institutional type homes. Consequently, it was amended in 1954 to provide federal funds for construction of public and voluntary nursing homes. States receiving these funds agreed to conduct inspection and regulation according to the provision in the Hill-Burton provision which reads:

See page 12 note card

> . . . the state health commissioner shall be responsible for administration of carrying out the law according to Act No. 88 of the Public Acts of 1943, and as amended in 1948, and in 1952 . . .[1]

It clearly defines that no person jointly or individually can establish or operate a home for four or more people without a license.

The accompanying note cards are those referred to in the report extract above. Notice that these cards have only a single entry. To avoid confusion when you get down to writing a report, place only one note on each card.

Notice how the last name of the author, Thomas, serves as a key to the bibliography card. Also, notice how the writer put headings on the note cards to suggest in which section of the report the note might be used: "Recognition of the problem," "Federal laws enabling nursing home licensing," and "Hill-Burton provision for enforcement." These headings serve as working outlines suggesting the possible organization of the report.

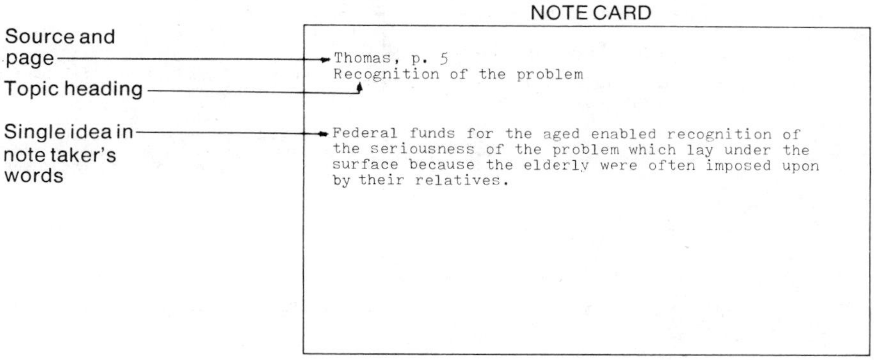

The quotation marks on the first note card from page 12 in Thomas indicate that this is a quotation, the exact words of the source. Always identify your quotations so you will know whether or not you will be required to identify the source in footnotes.

---

[1] William C. Thomas, *Nursing Homes and Public Policy* (Ithaca, N.Y.: Cornell University Press, 1969), p. 12.

NOTE CARD

Source and page → Thomas, p. 12

Topic heading → Hill-Burton provision for enforcement

Author's exact words (with quotation marks) → "... the state health commissioner shall be responsible for administration of carrying out the law according to Act No. 88 of the Public Acts of 1943, and as amended in 1948, and in 1952..."

The other note card from page 12 of Thomas concisely summarizes all the main ideas appearing on several pages. The note-card writer condensed the ideas on these pages to a short note. To do this, she used shortened forms of words and phrases that she can decipher, but which others might find difficult to interpret. The clearer statement of these ideas in the section extracted from the report illustrates how a brief note on a single card may be expanded.

NOTE CARD

Source and page → Thomas, p. 12

Topic heading → Federal laws enabling nursing home licensing

Summary in note taker's own words → Enactment by U.S. Cong. of Hill-Burton in '46 provided Federal funds for building public homes. '54 ammendment for private ones. It also provided for registration and inspection by state health commissioner, who may grant license to a person or group operating a home for four or more if it adheres to state and federal codes.

THE PARTS OF A REPORT

Before we get into the body or main part of the report, it is helpful to know all the parts that a complete report may include. These parts are listed below in their order of appearance. Items 1–8 constitute what is called the *front matter*. This is everything that precedes the report proper (items 9 and 10). Everything that follows the body of the report (items 11–14) is called *back matter*. These terms are also used with reference to the parts of a book. In this book, for example, the back matter consists of appendixes and an index.

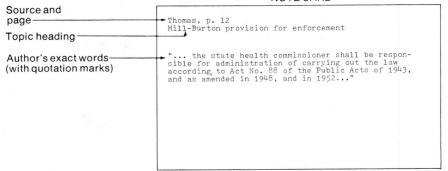

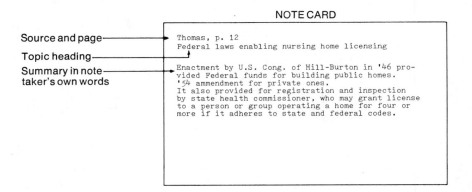

The parts of a report

1. *Cover*—The cover of a report, like the cover of a book, serves to protect the inner pages. Usually it presents the title and author. In short reports covers are usually dispensed with.
2. *Frontispiece*—Only formal bound reports include a frontispiece. The page faces the title page and it may be decorative or have a photograph or drawing related to the subject of the report.
3. *Title Page*—This page presents the title of the report, names of author(s), including their positions, and states where and when the report was prepared. It may also mention who authorized the report and for whom it was prepared.
4. *Letter of Authorization*—This document identifies the specific project, the question to be answered, or the problem to be solved. It authorizes the person named to make the investigation or to do the work, and permits others to help.
5. *Letter of Transmittal*—This is your record of sending the report. It may include your main conclusion or anything else you want to focus attention upon. In it you may also acknowledge the help you received in preparing the report.
6. *Preface or Foreword*—In the preface you state your purpose in writing the report and mention any limitations you may have set.
7. *Contents*—The table of contents lists the main sections of the report, with page numbers. It enables the reader to have a quick overview of the complete report and to locate certain material easily.
8. *List of Tables and Illustrations*—Like the table of contents, this lists the page numbers of the tables and graphic illustrations, to help the reader locate them easily. Such a list is given when there are a great many tables and illustrations.
9. *Abstract or Summary*—This is a brief condensation of the report, especially its main ideas, findings, conclusions, and recommendations. We have seen some examples and will see others when we take up the report proper.
10. *Body*—This is the report proper, the main part of the report. A long report may consist of any combination of the short technical reports studied in this textbook.
11. *Bibliography or References*—This is an alphabetical list of the books, journals, and other sources consulted in the preparation of the report. Cite the author for each (last name first), title, editor, city of publication, publisher, and date of copyright. For more exact and extensive directions refer to *The MLA Style Sheet* or *The Manual of Style*. Both of these are in most libraries.
12. *Appendix*—Here the writer puts any helpful supplementary material or any non-essential material that would disrupt the orderly development if it were part of the body of the report. Each appendix item must be clearly identified with a title. If the report has more than one appendix, each is labeled with a capital letter, for example *Appendix A, Appendix B,* etc.
13. *Glossary*—This is an alphabetized list of technical terms with their definitions. A glossary is very helpful when the intended reader may not easily understand the specialized terminology and, consequently, the report.

14. *Index*—Placed only in very long reports, an index consists of an alphabetized listing of key words to enable the reader to locate material in the report easily.

Headings of Subsections

When you are writing the body of a long report, it is a good idea to break up its text into sections. Doing this enables easier reading. Identify these sections, the main parts of the report, by using headings. Use headings of lower levels to identify the subdivisions of each section, much as a newspaper uses headlines for the main divisions and subdivisions of a story.

There are different ways of using report headings. The company for which you work usually has a preferred style. Whichever system you use, be sure to follow it consistently.

Here is a simple system to follow with an example of the heading for each level:

1. A section heading is in capital letters and is centered on the line. Like chapter titles in a book, this identifies the main sections of a report.

THIS IS A SECTION HEADING

2. The next level heading identifies the major divisions of the report. It consists of capital letters and is placed against the left margin.

THIS IS A MAJOR HEADING

3. The next level subheading is also placed against the left margin, but only the main words are capitalized. It is usually underlined.

<u>This Identifies a Subdivision of a Major Heading</u>

THE BODY OF A TECHNICAL REPORT

In Chapter 10 you saw how to write letters by first dividing them into three main sections. The body of a technical report is also divided into three main sections. These are: (1) beginning section, (2) developing and supporting section, and (3) concluding section. Knowing what each of these sections is supposed to do to achieve effective communication enables a report writer to decide on the kinds of details to include and how to organize these details.

Here is a typical outline of most technical reports:

I. Beginning Section

   A. Title
   B. Abstract
   C. Introduction

II. Developing and Supporting Section

III. Concluding Section

Beginning Section

To write reports fluently and effectively, you should try to understand the important functions each of these three main sections performs.

The title. A technical report begins with the title. Since it is your initial point of contact with your reader, it is the part of the report you will first rely on to establish reader contact. It also must identify what you are going to write about in a way that arouses the reader's interest.

To capture interest, a title must be appropriate to the type of reader for whom the technical communication is written. Following are examples of titles of articles intended for two different types of readers. Notice the difference in the wording.

For Restaurateurs	*For the General Reader*
"Brownies Are Profitable"	"Brownies Take the Cake"
"Using Bing Cherries for a Specialty of the House"	"Beautiful Big Bings"
"Improvements in Commercial Cooking"	"What's Going On in the Kitchen?"

Because technical reports and articles are aimed at experts, or at least at readers who are well versed in the subject matter, they often have lengthy titles. They are longer because they are expected to specifically identify the subject. Often, they are required to indicate the kind of technical report it is, its date of completion, and for whom it is intended. Following are some illustrations of report titles published by the U.S. Government Printing Office:

"Standard Specifications for Construction of Roads and Bridges on Federal Highway Projects"

"Selected Materials Concerning Future Ownership of the AEC's Gaseous Diffusion Plants"

"Characteristics of State Public Assistance Plans under the Social Security Act"

"Report of the Commission on Mortgage Interest Rate to the President of the United States and to the Congress"

Abstracts. We saw earlier that there are two main kinds of abstracts: (1) summary and (2) descriptive. The purpose of an abstract is to give the intended reader a quick indication as to whether he or she should closely study it, closely examine a part of it that is related to something of current importance, tell a subordinate to read it and give a review, or just have it filed away or discarded.

Most technical reports you write should have an abstract inserted and indented under the title. Each of the short reports used for illustration in this text will have a brief abstract. Long formal reports always have an abstract. Following is an abstract summarizing a formal report 83 pages long.

GEM STONES OF THE UNITED STATES [1]

ABSTRACT

Many semiprecious, but few precious, gem stones have been found in the United States. Beauty, durability, and rarity are the most important qualities of a precious gem. Gem stones are distinguished by their physical properties: color, crystal form, cleavage, parting, hardness, specific gravity, luster, index of refraction, transparency, and dispersion. Gems are named for their color, type locality, outstanding physical property, or persons. The most popular gem cuts are the cabochon, rose, brilliant, step, and mixed. The carat, one-fifth of a gram or 200 milligrams, is the unit of weight measurement. The color of four popular gems may be changed by heat treatment or dyeing. Only the ruby, sapphire, spinel, emerald, rutile, and quartz of gem quality have been synthesized. The best quality of assembled stones are the doublet and triplet. Most gem stones are found in alluvial gravels and igneous rocks, especially granite and pegmatite deposits.

The descriptive abstract below from the U.S. Department of Transportation [2] illustrates clearly how much time and work these report summaries can save you. It briefly describes the content of a 76-page report about the research conducted to determine the most useful truck-top markings for security. Imagine how much work you would have to do, to say nothing of the cost, if you had to do your own research.

TRUCK-TOP MARKINGS FOR VISUAL IDENTIFICATION

Describes a project initiated by the Department of Transportation to determine the most useful size, shape, location, and color for security-related truck-top markings. Guidelines are presented for the appearance of markings and for suitable materials to be used. 1973. 76 p.

Here is an abstract of another report resulting from a long study:

THE EFFECT OF GROYNES ON ERODED BEACHES

Laboratory tests are described, in which the effect of impermeable groynes on an eroded beach was studied. A beach was allowed to reach equilibrium

---

[1] Dorothy M. Schlegel, "Gem Stones of the United States," Geological Survey Bulletin 1042-G (Washington, D.C.: U.S. Government Printing Office, 1966), p. 203.

[2] "Selected Government Publications Bulletin" (Washington, D.C.: U.S. Government Printing Office, March, 1976), p. 5.

for a particular wave climate and supply of littoral material. The foreshore was then manually eroded, and the beach allowed to return to equilibrium with and without groynes. It was found that the presence of groynes increased the rate of accretion but did not significantly build up the inshore beach beyond the stable levels. Bed levels seaward of the groynes were increased.[3]

Introduction. The introduction of a technical report names the specific topic and tells the main thing the writer wants to say about it. This is done in what is often called the thesis sentence, the topic sentence, or the subject sentence. In a short report, the introduction usually has one or more sentences explaining further what the topic is, what it does, or how it is used. In longer technical reports, you may have to use several paragraphs instead of a few sentences to do this same job.

The introduction of a technical report may have to provide a reader with some or all of the following information:

1. Background information about the history of the subject and main idea and the reason for the continued interest in this report.
2. Principles and technical theories needed for understanding the ideas in the report.

Background Information. Often a reader can't understand what a writer is saying in a technical report because he or she does not have enough background experience or information. This is not always just for lack of formal education. The reader might be the president of the company who has been occupied with other matters and unable to keep up with all of the details of the business. He or she must be given the history and other explanations needed to understand what is said in the report.

The historical background you give may trace the origin and development of an object, person, or idea discussed in your report. It may fill the readers in with details about when and where the subject or main idea originated, who developed it, the main causes for its growth and development, and its status today. For example, for readers fully to understand a certain technical problem, you may have to tell them something about the history of the person or company having the problem. Doing this may enable them to accept the solutions or recommendations you will offer later. Doesn't a physician have to have the patient's case history before treating an illness effectively?

Definitions are often an essential part of your report introduction. New inventions, discoveries, and information are added constantly to commercial and industrial technology. It's almost impossible for a person to keep up with all of them. Many of these new terms, along with the more common ones, may mean different things at different times. Therefore,

[3] W. A. Price and K. W. Tomlinson, "The Effect of Groynes on Eroded Beaches" (New York: American Society of Civil Engineers, 1971) pp. 1053–58.

you will have to be sure that the readers understand exactly what your report is about.

Your readers will have to understand the technical principles before they can interact as you think they should with what you say. You may have to provide them with explanations of the principles or basic laws, theories, equations, or formulas causing something to be able to exist or happen. Unless the readers understand that some gases are lighter than air, they might not be able to understand why dirigibles stay aloft over a football stadium. A city, state, or federal law may have to be explained to enable the reader to understand why doing business a certain way is illegal. The law of supply and demand is an important economic principle that you may have to explain so your reader can grasp what you say later. You may have to tell readers that electricity creates heat when it encounters resistance, so that the more inefficient a conductor is for transmitting power, the better it produces electric heat.

As an illustration, here is the introduction from "Gem Stones of the United States" by Dorothy M. Schlegel. This introduction is mainly concerned with giving the readers needed definitions and explanations of the technical principles.

INTRODUCTION

Gem stones generally are divided into two categories: precious and semiprecious. A precious gem stone has beauty, durability, and rarity, whereas a semiprecious gem stone has only one or two of these qualities. The diamond, emerald, ruby, and sapphire are considered precious gems. Some opal is precious, but most varieties are semiprecious.

The beauty of a gem stone is determined by personal taste. In ancient times man preferred brightly colored, translucent or opaque stones. Today he prefers evenly tinted, transparent stones. The desired hues are blue, rose, green, and true canary yellow in the diamond; pigeon-blood red in the ruby; cornflower blue in the sapphire; and grass green in the emerald. Most diamonds, however, are colorless.

The durability of a gem stone depends upon its hardness and lack of ready cleavage. A gem must be sufficiently hard to resist abrasion by objects normally found in everyday life and by dust. It should also resist the chemicals with which it comes in contact. Cleavability is the tendency of certain gems to split in one direction more readily than in another.

Rarity is one of the most important factors in establishing the price of a gem stone. Such gems as the diamond and ruby are rare, in addition to being beautiful and durable, and therefore are very expensive. Although the deep red pyrope garnet closely resembles the ruby in color, there is no comparison in expense and popularity.[4]

---

[4] Schlegel, p. 204.

Developing and Supporting Section The developing and supporting section of a technical report amplifies and adapts the main ideas to evoke a certain type of response from an intended type of reader. You, like any technical writer, will spend most of your time writing this part of the report.

Most of the chapters in this book tell about the kinds of ideas that go into this section and how to develop the supporting paragraphs in it. Chapters 1, 4, 5, 6, 7, and 8 discuss reader adaptation and how you can amplify your ideas to achieve the kind of communication your report intends.

In the chapters following this one, you will learn how the developing and supporting section of a status report amplifies the description of the condition of its subject. You will see how the section of a process or procedural report (Chapter 14) clarifies the sequence of steps involved, and how in an interpretive report (Chapter 15) it clarifies the way an investigation was conducted, and its findings and conclusions.

This developing and supporting section of a technical report distinguishes it from the other types of reports. Therefore, it is important that you learn about the different kinds of ideas and the different ways of organizing them to get the response you want.

Concluding Section The concluding section gives,the final touches to the report by restating the main ideas that the writer wants the readers to understand and remember. It may aslo point out and develop any relationships these ideas may have to future happenings.

The concluding section of any report should end strongly. A weak ending gives the impression that the writer has been scraping the bottom of the barrel for something to say. A busy reader does not have time to waste on second-rate ideas; consequently, he becomes very irritated at this kind of writing. Nor should technical writing of any kind begin or end with an apology; the reader will find the weaknesses without your pointing them out. Rather than apologize for a report's weaknesses, correct them before writing it.

BIBLIOGRAPHY AND FOOTNOTES

As the writer of a technical report, you are accountable to your readers for the reliability of your information and the qualifications of your sources. You should be able to provide whatever documentation is needed to support what you write. The documentation in a technical report consists of two main kinds of references: (1) bibliography and (2) footnotes. The entries on note cards from which you will prepare a bibliography, identifying the sources used in a report, were discussed earlier in this chapter. Only one entry is placed on each of these cards. When the report is finished, you simply arrange them in alphabetical order and type them on a page headed "Bibliography."

Footnotes are references explaining an idea or identifying the source of one. Often they are placed at the bottom (or "foot") of the page where the reference number appears in the report, separated from the body of the page by a short horizontal line. The numbers may start with "1" on each page or they may be numbered consecutively through the report. Another way of presenting footnotes is to list all of them in numerical order on a separate page at the end of the report.

A third method of documentation widely used in scientific and technical papers is to present a numbered bibliography, and insert the proper reference number directly after the source when it is mentioned in the report. In such a report, you would see, for example:

> Enrichment of white flour to restore vitamins lost in milling has been recommended by many workers, and different forms and combinations have been proposed. Dawson and Martin (7) recommended the addition of dried yeast as a supplementary source of vitamins, whereas Robertson (11) claimed that bread made from white flour fortified with thiamine, riboflavin, niacin, and iron is the best from the standpoint of nutrition. Menden and Cremer (3) have reviewed the method and levels of enrichment in various countries.

To find out what source the number refers to, the reader consults the corresponding numbered item in the bibliography. Obviously, this is a time-saver both for the reader and the writer of the report.

Source footnotes have the following parts: (1) author (first name first), (2) title, (3) facts of publication (name and location of publisher and date), and (4) page number(s). All of this information must be included the first time you footnote a source. However, in subsequent footnote references to the same source, the entry is greatly shortened.

The list below includes examples of "first-time" footnotes, which you may use as a guide. You may get explanations of other conventions by referring to any of several reference books such as *A Manual of Style*, published by the University of Chicago Press, or *The MLA Style Sheet*, by the Modern Language Association of America. Both are available in most bookstores. The following examples are based on *The MLA Style Sheet*.

As stated earlier in this chapter, the information for a footnote is keyed by the last name of the author on each note card to the bibliography card. The first three illustrations below will enable you to see easily the difference between a bibliography and two kinds of footnote entries.

1. Bibliography entry:
   Mossman, Frank H. *Principles of Urban Transportation.* Buffalo, N.Y.: Western Reserve University Press, 1951.
2. First footnote entry:
   Frank H. Mossman, *Principles of Urban Transportation* (Buffalo, N.Y.: Western Reserve University Press, 1951), p. 25.
3. Subsequent references:
   Mossman, p. 25.

4. Standard footnote reference to a book with more than one author:
   Tom E. Wirkus and Harold P. Erickson, *Communication and the Technical Man* (Englewood Cliffs, N.J.: Prentice-Hall, Inc., 1972), p. 33.
5. Footnote reference to a book assembled by an editor:
   Herman A. Estrin, ed., *Technical and Professional Writing* (New York: Harcourt Brace Jovanovich, 1963), p. 18.
6. Footnote reference to an article from a magazine or journal:
   Michael Lipman, "How to Build Sales with House Organs," *Business Management*, 15 (November 1961), p. 102.
7. Subsequent magazine article references:
   Lipman, p. 78.
8. Footnote reference to a newspaper article:
   Kathy Warbelow, "Can a Big City Cop Find Happiness as a Small Town Lawman?" *Detroit Free Press*, April 11, 1976, p. 12.
9. Reference to an encyclopedia or other alphabetically arranged reference book:
   "Diseases of Cattle," *The Encyclopedia Americana*. (Since arrangement is alphabetical, page and volume numbers are not needed.)

The Latin abbreviations *ibid.* for "the same" and *op. cit.* for "in the work cited" are seldom used today, but you may come across them in older references.

Since a bibliography lists all books and articles you used in preparing the report in alphabetical order, the last name of the author is given first. This is different from footnote style, in which the first name of the author is placed first. The bibliography, as shown below, gives the important facts of publication. This enables your reader to go to your source to get additional information about something you said or to check whether you quoted or interpreted the information accurately from your source.

Compare the following bibliographical entries with the preceding footnote forms. Be sure to notice also how the second and third lines are handled. In the footnote they are brought out to the left margin, but in the bibliography they are indented.

"Diseases of Cattle," *The Encyclopedia Americana*, 1972 ed.

Estrin, Herman A., ed. *Technical and Professional Writing:* A Practical Anthology. New York: Harcourt Brace Jovanovich, 1963. (Subtitle was not given in footnote.)

Lipman, Michael. "How to Build Sales with House Organs," *Business Management*, 15 (November 1961), 99–110.

Mossman, Frank H. *Principles of Urban Transportation*. Buffalo, N.Y.: Western University Press, 1951.

Warbelow, Kathy. "Can a Big City Cop Find Happiness as a Small Town Lawman?" *Detroit Free Press*, Sunday, April 11, 1976, p. 25, col. 4.

Wirkus, Tom E., and Harold P. Erickson. *Communication and the Technical Man*. Englewood Cliffs, N.J.: Prentice-Hall, Inc., 1972. (Only last name of first author is given first for alphabetizing.)

Application 12–1 Select any five of the follownig topics, and in your library find at least three sources not written by the same author. Write a bibliography for all five.

1. Industrial pollution of the Great Lakes.
2. Glass manufacturing in Venice, Italy.
3. Shoe manufacturing in Massachusetts.
4. Building the Suez Canal.
5. How books are bound.
6. Gases in rocks.
7. Granite quarrying in Georgia.
8. Nut farming in California.

Application 12–2 Brainstorm a specific topic and main idea suitable for a technical report by leafing through each of the following: (1) subject card catalogue in your your library, (2) *Reader's Guide to Periodical Literature,* (3) a technical journal index, (4) a technical abstract index. Write the topics and main ideas and opposite each tell which index suggested it.

Application 12–3 Using two of the main ideas in the preceding application, find two sources from which you can derive worthwhile information about your main idea. Write three note cards for each of the two main ideas. Be sure to head your notes properly. You may want to key them in to the bibliography cards you did for Application 12–1.

Application 12–4 Select any two of the articles you found in the *Reader's Guide* or one of the other indexes. After reading each carefully, write a descriptive abstract for one and a summary abstract for the other.

Application 12–5 Indicate whether the following subjects could be written about more effectively (a) as an informative report, or (b) as an interpretive report.

1. Tracing the invention and development of the ballpoint pen.
2. Giving the causes you think should be checked to account for an employee's injury.
3. Explaining how a person should go about getting a certain job.
4. Telling the service manager of an automobile agency what you think is causing your car to malfunction.
5. Reporting the number of trucks in good working condition on a certain day.
6. Explaining what you think caused a loss in profit for a small business you manage, along with solutions for the problem.
7. Giving your judgment of the durability of certain pieces of work clothing you want to rent for your helpers.
8. Setting forth possible solutions for a problem should it ever become a reality.

Application 12–6 Identify the general kind of technical report that each of the following describes by printing "A" before each one that you think is an informative report or "B" before each one that you think is an interpretive report.

1. A report presenting practical methods for sanitizing home laundry equipment. This report discusses the uses of several types of disinfectants found effective in killing bacteria during laundering.
2. This report explains the uses of DDT in U.S. agricultural applications and makes predictions as to the effect it will have on future farm production.
3. A report that warns consumers about the false and misleading advertising of "bulk" meat bargains for home freezers. It tells the consumer about the characteristics and tactics of misleading advertisements.
4. This pamphlet report provides helpful instructions for the victims of an earthquake. It tells them what they should do during and after the shaking to protect themselves.
5. A claim report describing the results of a fire; it also explains what may have been the cause for the fire based on two empty gasoline cans found in the stock room.
6. A technical report describing a television X-ray image storage system designed to reduce radiation exposure.
7. A report explaining what oil shale is. It discusses its history, how it is formed, where deposits are found, and its uses.
8. This report discusses the trends in labor costs for the coming year and makes recommendations for increasing production to accommodate the expected labor cost increases.
9. A technical report that provides information on the digital electronic computers installed in the central office of a large corporation. It also contains tables giving the numbers and types of computers by location, model number, and manufacturer.
10. A report that presents a summary of the major changes in salary and supplementary (fringe) benefits that have taken place in production departments of the automotive industry in Detroit, Michigan, during 1972.

Application 12–7 Write a two- or three-page informational report about the parking facilities at the college you attend, a shopping center in your area, or a large manufacturing plant in your city. Give the names by which the various parking areas are identified, where they are located, the kinds of vehicles required to park there, and the capacity of each area. You might also discuss the procedures for securing permits to park in these areas and the stickers that must be placed on the cars authorized to be parked in them.

13

Writing status and progress reports

what status reports are . . . how they differ . . . their purposes . . . kinds of status reports . . . how to develop and organize them . . . kinds of progress reports: (1) preliminary, (2) interim, (3) final . . . ways of organizing and developing progress reports . . . examples of status and progress reports

To write a good technical report, you first should know your purpose. In several of the earlier chapters in this book we discussed a writer's purpose as the response intended from the reader. You should know whether you intend to inform or to persuade your reader. Knowing this will help you decide the kind of report you should write.

The following kinds of informative and interpretive reports are explained and illustrated in this and the next two chapters:

Informative Reports:

1. Status
2. Progress
3. Process or procedural
4. Historical

Interpretive Reports:

1. Evaluation
2. Causal
3. Proposals
4. Feasibility

STATUS REPORTS

A status report is informative, not interpretive. Its main function is to describe a material, a device, a place, a person, or an organization by telling what it is, what it looks like, or what condition it is in at a certain time. It may be a description of a railroad marshaling yard, the place where railroad cars are arranged in the order of their destination; it may be a description of an organization, e.g., General Motors; a device, e.g., a lawn mower, a person, the president or the sales manager of a company; or it may define and describe a material, e.g., polyester resin or a plastic made from it. In other words, informational status reports communicate the status quo of a technical subject.

Claim or accident reports, when they are mainly concerned with conveying the nature and the extent of the damages resulting from a mishap, not with what caused it or how it occurred, are more specific kinds of status reports. These reports are written during or shortly after an inspection is made of the object or person involved in the happening. Damage reports, police reports, fire reports, and physical examination reports are other kinds of status reports related to claim or accident reports.

Work reports and cost estimates are status reports that describe the work that has to be done on a certain job along with the cost of the material and labor. They do not, however, describe how the work is to be done; that is done by another kind of report, the process report. A work report may include specifications of the ingredients or components that will be used.

Although there are many different kinds of status reports, most of them contain much of the same type of information, arranged in somewhat the same order. Following is a very common format for writing one. True, this is not the only way to develop a status report; however, it is a good way. Use it until you can develop or find a more suitable one.

Beginning Section

A. *Title*

1. Identifies as specifically as is necessary the subject of the report, i.e., the name of the material, device, project, place, organization, or person about which the report is written.
2. Indicates the kind of status report intended, e.g., claim report, accident report, work report, etc.
3. If necessary, it indicates the date of the report.
4. May state the main idea of the whole report.

B. *Abstract*

Gives a concise summary of the main ideas in the whole report: the specific subject, the scope and purpose of the investigation, and the main findings. It may contain only one sentence or several paragraphs, depending upon the length of the original report. It usually is written when the report is finished. Some abstracts describe the report rather than summarize it.

Writing status and progress reports

C. *Introduction*

1. The first paragraph contains the thesis or purpose statement, which identifies the specific subject and the main idea expressed about it in the report. It also may tell who wanted the report done and why.
2. Background information gives the reader definitions or historical background tracing the origin and development of the subject. A brief explanation of the meaning of terms or principles of physics, chemistry, physiology, economics, etc., that enable the subject to exist or function may be given here.

Developing and Supporting Section

A. *Overall Description* of the external appearance of the subject, giving details related to the main idea and to the objective of the report. This may include any or all of the following: size, shape, design, color, dimensions (weight, height, depth, width, etc.), or texture.

B. *Structural Description of Main Ingredients, Parts, or Components* of the subject, especially those related to the main idea of the report. If the main components are moving parts, it may describe how they function. If the subject has no moving parts, e.g., a knife or a hammer, this section may describe the main use or purpose.

C. *Functional Description* of the whole device or organization in motion as a unit, working together. If the subject is stationary or has no working parts, its uses or purposes are given.

D. *Helpful Drawings, Diagrams, or Photographs* may be used to make the description more vivid.

Concluding Section

A. A summary of the main ideas of the report, especially those related to the operational and functional applications of the subject.

B. A restatement of the values of the subject to the reader, especially those related to its improvements or advantages.

You can see from the amount of description needed in the developing and supporting section of a status report how essential it is. At this point, therefore, it would be a good idea for you to go back and review carefully the discussion of description in Chapter 5.

Notice how the abstract in the following status report is set apart and clearly identified by the word "ABSTRACT" just below the title. The student's drawings are very helpful, even though they are not the work of a professional.

Beginning Section

Title

Avoiding Transmission Leaks by Pneumatic and Hydraulic Testing of Cases

ABSTRACT

The hydraulic tester for transmission leaks provides a reliable method for detecting transmission defects before installation. It helps increase the speed of automobile and truck production along with their durability.

The hydraulic tester enables two tests to be given simultaneously to determine whether or not a leak exists in a transmission, to locate it if it does, and to determine its size. It does this by testing the case in two ways before it is assembled around a transmission.

Introduction: purpose statement

The hydraulic tester functions pneumatically as well as hydraulically. It tests a transmission case pneumatically by inserting air pressure into it and hydraulically when the case is submerged into its water tank. In other words, it operates on the same principles as that applied when finding a small leak in rubber tubes by filling them with air and dunking them into water.

Background: principle of physics

This, water tester for transmission cases was first developed at the beginning of World War II, around 1940. It was needed to speed up the production of vast numbers of durable war vehicles, tanks, trucks, jeeps, and many others required to travel over rugged terrain and not break down. Before that time, these cases were carefully tested by air and close observation alone. However, this was not good enough for rapid production of reliable vehicles.

Historical

The hydraulic tester used to locate leaks in transmission cases is 12′ wide, 14′ long, and 8′ high. It rests on a conical shaped cement pad for easy drainage. On its left side is the electrical control box. In front of it is its gauge board. Behind this gauge board are the hydraulic and pneumatic components. Its 100-gallon water tank consists of two compartments; one is empty and the other is full of water. The water is transferred from one compartment to the other when the operator wants to change the level of the water.

Developing and Supporting Section

Overall description
External overall description

Detailed description of main component

In the accompanying drawings, Figure 13.1 is a front view of a hydraulic tester. Figure 13.2 is a right side view showing the design and components.

It takes 30 seconds to complete one cycle of the hydraulic tester. A transmission case is put into the empty tank and set in a frame to hold it firm. When the operator pushes the two front controls, the machine starts its testing cycle. The machined holes in the transmission case are sealed with rubber plugs inserted at the end of small hydraulic shafts. When the machine starts, these shafts slide in and out. Air is then introduced into the transmission case; 14 P.S.I. air pressure is inserted into each one. The pressure is indicated on the round gauge below the gauge board. If there is a leak in the case, one of the big square gauges will indicate the air pressure loss, and red lights on the gauge board will start flashing.

Functional description (Whole device in motion)

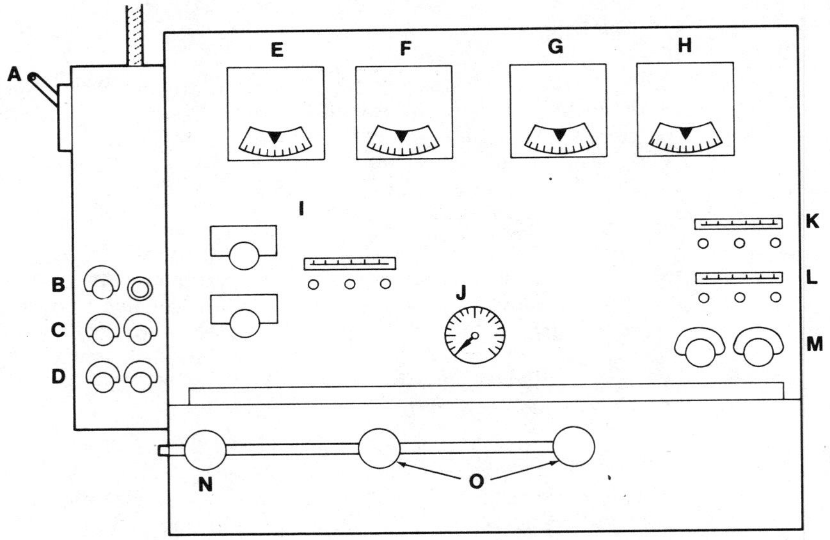

A	Main Switch	**I**	Reset Control
B	On/Off Control	**J**	Pressure Gauge
C	Manual and Automatic Air Control	**K**	Cycling Reject Lights
D	Water Level Control	**L**	Accept Light
E,F,G,H	Leak Locating Gauges	**M**	Reset Buttons
		N	Uncycling Control
		O	Cycling Controls

***Figure** 13.1*
Front view of hydraulic tester.

***Figure** 13.2*
Right side view of hydraulic tester.

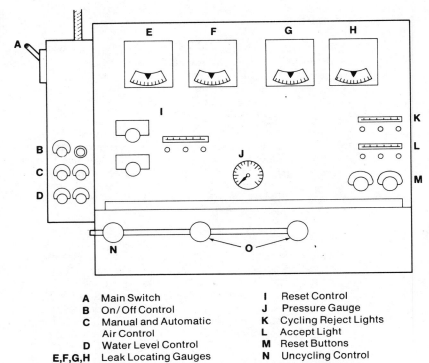

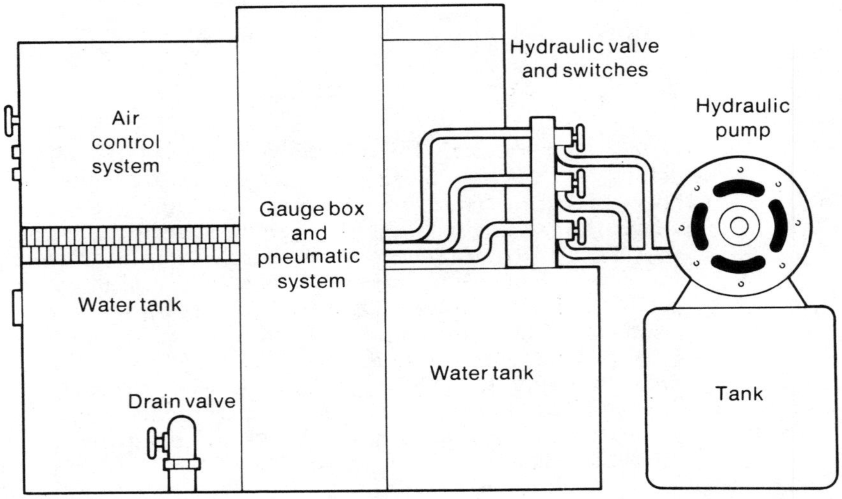

Immediately after the loss of pressure is detected, the water is raised over the case. The air still inside will continue to escape, causing bubbles to appear in the water. By following the bubbles, the operator will be able to pinpoint the exact location of the leak.

Concluding Section (Summary and restatement)

These hydraulic testers enabled very efficient testing of war vehicles during World War II, and are still being used by the large automobile and truck manufacturers. They greatly increase the safety features and the durability of American-made cars. An average of 200 parts are run through the typical water tester during each shift at one large plant. It is reported that about one-third of the cases are discarded because they have excessively large leaks. They are usually resmelted and remade into new ones.

Status reports are so common that industrial concerns and other organizations often prepare blank forms for their employees to fill out, thus saving work in preparing them and insuring that the necessary information will be included.

PROGRESS REPORTS

A progress report is just another kind of status report. The status report, however, is a single report describing the condition or the appearance of something at a given time, whereas a progress report is one of a series of reports, each describing the status of something at a given time while it is undergoing construction or change. As the name "progress" report implies, it is a report of a subject that is in the process of being completed.

A progress report may discuss anything that is undergoing change. It may be about a patient undergoing medical treatment, the periodic results of some kind of investigation, or the work completed during a certain period on a project under construction, perhaps a bridge or a building.

Obviously, therefore, the reader of a progress report is usually a person interested in the project. He or she may be someone who will benefit from it. Often, the reader is a person who is responsible for knowing the status of a project so that he can make decisions concerning it. Therefore, the reader may have to know what was or was not achieved, the problems that were not resolved as well as those that were. The report also may identify problems that may arise in the future so that steps may be taken to overcome them.

There are four main kinds of progress report:

1. The preliminary progress report.
2. The initial progress report.
3. The interim progress report.
4. The final (or completion) progress report.

The preliminary progress report is written before the actual work on the project is begun; therefore, it is the first report. Its main purpose is to indicate to what degree the supplies, equipment, personnel, and other necessary elements have been gotten ready in preparation for starting the project.

The initial progress report is the first report prepared after the work has started. Therefore, it is very similar to the interim progress report. This text will give a detailed discussion of the interim progress report, not the initial progress report, since both contain the same kind of information.

The interim progress report (or reports) is written in between the preliminary (or initial) report and the final progress report. Usually, there are several interim progress reports, each indicating what has been achieved since the last report.

The final (or completion) progress report, of course, is prepared after the project has been completed.

A preliminary progress report, as we have noted, is the first of a series of progress reports intended to keep a reader regularly informed about extent to which something has moved toward its completion. It is prepared before the work is started; therefore, it cannot tell what portion of the project has been done nor can it discuss problems encountered while the main project is in progress. Its main purpose is to discuss the preparations made before hand to enable the project to start on schedule. Also, it usually identifies and discusses any problems that may have occurred while preparations were being made, how they were handled, and whether or not the work will begin on schedule.

Following is a common pattern which, with slight changes, you can use as a guide in writing any of the listed progress reports on the preceding page. Although not all of the details in the following format will always be needed, any progress report will usually contain most of them. After studying and following the suggested pattern, you will be much better able to improvise and adapt it to any of the progress reports discussed.

Beginning Section

A. *Title*

May indicate the kind of progress report it is (preliminary, interim, etc.), the specific name of the project, and the time period the report covers.

B. *Abstract*

Briefly summarizes what was accomplished in the report's time period, whether or not the work is progressing as scheduled, the main problems encountered, and what was done to work around or solve them.

C. *Introduction*

1. Contains the purpose statement, identifying the project, and briefly indicates the main things done since the last report and whether or not the work is on schedule.

2. May state who authorized the project and why it was authorized, its main purpose.
3. May give background information for the project: (a) define important terms or ideas that must be understood and (b) supply any needed history related to the origin and development of the project.

Developing and Supporting Section

The main differences in progress reports occur in this section; therefore, the suggested patterns for their development are shown under separate headings.

A. *Preliminary*

1. Preliminary Preparations
Discusses the preliminary preparations, especially the securing of needed personnel, materials, and equipment.
2. Problems
Discusses problems encountered while making the preliminary preparations and how they were resolved, especially those related to essential personnel, materials, and equipment.

B. *Interim*

1. Progress Achieved (or Lack of It)
   a. A detailed, accurate, and clearly presented itemization or discussion of the work completed since the last report.
   b. Identification and discussion of the problems that may have occurred during this report period, how they were resolved, or why they were not resolved.
   c. Identification and discussion of the changes in personnel, materials, or apparatus that enabled increased progress.
2. Work Still To Be Completed
   a. Discussion of the work that may still have to be completed.
   b. Identification and discussion of any anticipated problems.

C. *Final*

1. Completion of the Project
Discusses in detail the work done on the project since the last Interim Report.
2. Problems
Discusses the problems encountered since the last Interim Report, especially those related to the job completion.

Use helpful tables, charts, graphs, and/or drawings in any of these progress reports to reduce and make clear your explanations.

Concluding Section

A. Summarizes the accomplishments, the main problems encountered, and the way they were resolved.
B. States whether or not the project is on schedule. If not on schedule, it explains why, and may offer suggestions as to what can be done to help it along.

Following are examples of the progress reports discussed in this chapter. The student writer of the preliminary progress report relies heavily on drawings to explain the progress made in remodeling a basement.

Beginning Section

Preliminary Progress Report on Remodeling A Basement—January 7, 1977

ABSTRACT

After jointly deciding to remodel our present home, instead of buying another, for more comfortable accommodations, we drew plans and now have all the tools and materials, except the copper piping. We intend to start remodeling the basement into four areas: a den, recreation room, workshop and a laundry-sewing room. We are ready to begin work on January 8, 1977.

Introduction: purpose statement

Identifies the project and its purpose

Shortly after deciding that it was necessary for the writer of this report to remodel the basement of his home himself, his family immediately began to help him to do the necessary preliminary preparations. They decided that it was less expensive to remodel the lower level than to add rooms to the upper floors, and doing it would not increase the tax base very much. It would also enable the maximum use of an additional one thousand square feet of space.

Background: history

The house we, three teen-agers and two adults, own is a three bedroom bungalow located on Haggerty Road, Livonia, Michigan. The house was purchased in 1968, when the children were satisfied with games and outdoor sports. Now they are older and require more sophisticated kinds of entertainment. Therefore, we need space for them to have their social activities, club meetings, parties, and dances. A remodeled basement is ideal for these activities.

Developing and Supporting Section

The family prepared a preliminary drawing which called for dividing the basement into four main areas and a small storage room. Figure 13.3 is a copy of this drawing.

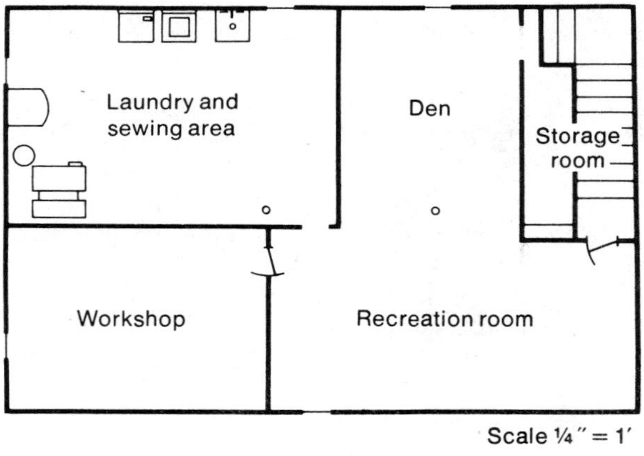

Figure 13.3
Preliminary drawing for remodeling of basement.

The drawing clearly shows that the remodeling requires the basement to be divided into a den, a recreation room, a workshop, and a combination laundry-sewing room. The intended den will be 12′ × 25′ and will have a game table and a ping pong table. The ping pong table will be portable and can be easily folded and stored against the wall.

Preliminary preparations

The drawing shows the workshop to be 12′ × 12′, a size convenient for any type of house maintenance work. The laundry-sewing room will be a little larger than the workshop; it will be 12′ × 18′. In addition to the automatic washer and dryer space, it will have room for a built-in sewing center. The sewing area will include a sewing machine, ironing board, full-length mirror, storage shelves for material, and a clothes closet.

Since the remodeling of the basement is to be done by the writer with the help of the other family members, gathering the tools, equipment, and needed building supplies was started immediately after completing the drawing. Here are the main tools that were purchased or rented: circular hand saw with a 7″ blade, a sabre saw, claw hammer, screw drivers, wire cutters, level, carpenter's square, tape measure, chalk string to insure that the partitions are in direct line with the ceiling and the existing walls. All of these items are on the site of the remodeling.

The required building materials and supplies have been ordered, and all of them have been delivered with the exception of the copper piping for the water line running to the sink and clothes washer. These will be delivered next week, long before they are needed. The following have already been delivered: sixty-five (65) 2″ × 4″ × 8′ boards, twenty-two (22) sheets of plaster board, five (5) pounds of sixpenny nails, three (3) pounds of plasterboard nails, two rolls of seam tape for plastering the wallboard seams, twenty-five (25) pound bag of finishing plaster, three hundred (300) feet of lomax 12-2 electrical wire, twelve (12) outlet boxes for switches and plug recepticles, seven (7) three-hole plugs with box, and four (4) circuit breakers.

Problems

The main problems encountered in making the preliminary preparations for the project concerned agreeing as to which plan provided the most convenient arrangement of the rooms. We didn't want everyone to have to walk through the workshop to get to the recreation room. Concern was expressed about the location of fixed structures such as ceiling support posts, the furnace, and the water heater. The location of heating ducts, window, and doors also gave us some trouble. All of these problems were resolved to everyone's satisfaction, and the final plans are as indicated on the drawing.

To insure that all of the preliminary preparations have been made, a recheck of the materials and equipment has been completed in preparation to begin the remodeling tomorrow morning, January 8, 1977. This is one week ahead of schedule. This was made possible by our purchasing most of the supplies, except the copper pipes, from local suppliers.

Concluding Section

Next is the first and only interim progress report explaining the progress made in remodeling the basement since the preceding preliminary report.

Beginning Section

Interim Progress Report on Remodeling A Basement—February 15, 1977

ABSTRACT

The project of remodeling our basement is moving along on schedule. Warped joists and the uneveness of the cement floor caused us some difficulty. This was corrected by means of shims and wedges as shown on the drawings.

Introduction: purpose statement

The basement remodeling started by the writer of this report on January 8, 1977, is moving along on schedule. Although the project was started one week ahead of schedule, time was lost because the writer, who is doing most of the work himself, had to attend a convention in New York.

Relationship to preceding report

The thoroughness with which the preliminary preparations to remodel the basement were made has greatly helped. The specifications shown in the preliminary report and the plans accompanying it were clear enough to enable the writer's teen-age sons, John and Henry, to rearrange and move the items in the basement to the garage for temporary storage.

Background information

As stated in the preliminary report, the purpose of the remodeling is to provide more comfortable living area for the family. To do this, the basement is to be divided into a den, a recreation room, a workshop, and a combination laundry-sewing room. These rooms will be equipped with the necessary furnishings for the intended activity.

Developing and Supporting Section

Progress achieved

The first thing the writer did on returning from the convention was to mark off on the basement floor with chalk where the partitions were to be laid. Each wall joist had to be measured separately before the partitions could be nailed together.

Progress reports

Today, all of the partitions are up, and the basement is divided as shown on the complete drawing sent with the preliminary report. The hardest parts to build were the doors. However, these also are all installed in their respective partitions.

Our main problem in erecting these partitions was experienced in getting them level. This was caused by the uneven basement floor and walls that were not square. Consequently, much time and effort was spent in making the necessary adjustments. To make these corrections, we had to use wedges cut from 2 × 4's as shims to make the partitions level with the floor and existing walls. This was done after the partitions were set in place along the chalk lines marked along the floor and walls. When the partitions were level, they were nailed down.

Problems encountered

Installing the door frames was also a time-consuming problem. When we started to install them, it was discovered that the wall joists at the doors opening to the storage areas and the workshop were warped. After a great deal of work, we solved this problem also with the use of shims.

Figure 13.4 is a portion of a wall frame section showing how the correction was made by the use of shims. This had to be done because the basement floor apparently settled sometime after it was poured, causing it to become uneven.

Figure 13.4
Adjustments for uneven floor and walls.

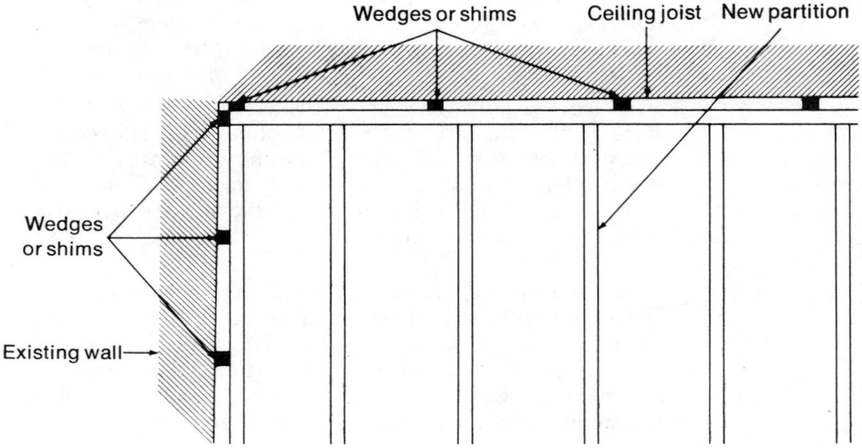

Figure 13.5 shows where shims were applied to correct the warped 2 × 4 in the door frame.

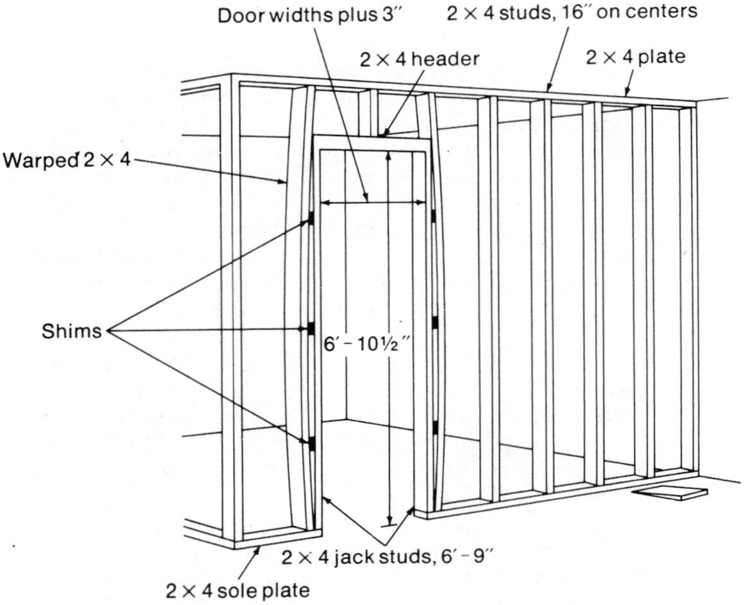

Figure 13.5
Correction of warped 2 × 4 in a door frame by means of shims.

Now that the partitions are up, we will start right away to decide on the location of the electrical outlets and switches so that the wiring can be installed. When this is done, the drawings for the wiring can be made to enable a licensed electrician, already contracted, to start work.

Work still to be completed and the prognosis

When we have done all of the electrical work, we can start laying the pipe for needed plumbing. The copper pipes, which were not here when the preliminary report was written, have arrived and all of the needed plumbing supplies have been delivered. No problems are expected in either the wiring or plumbing.

Anticipated problems

As soon as the wiring and plumbing have been completed, we will decorate the interior. Since the wiring and plumbing should be completed by February 19, installing the wallboard and painting will be finished probably by February 25. By March 15, the whole project, including the interior decorating, should be ready for our enjoyment.

Concluding Section

The final progress report on the basement remodeling project follows.

Final Progress Report on Remodeling a Basement, March 15, 1977

Beginning Section

ABSTRACT

The project of remodeling our basement has been completed on schedule, and excepting for two pieces of furniture, the den and recreation rooms are furnished. The project was facilitated by the careful plans drawn before we started working.

Introduction: purpose statement

In the Interim Progress Report of February 15, 1976, we projected completion of the basement remodeling by March 15, 1976. We are happy to report that with the exception of an undelivered ping pong table, the project is finished, and we are enjoying the results of our work.

Background information

As stated in the last interim report, the remodeling still needing completion at that time was installing the wiring, plumbing, and wallboard, mainly. The finishing work, including the painting and interior decorating, also had to be done. As we got closer to completing the project, the work remaining became less difficult and more enjoyable.

Developing and Supporting Section

Work done since the last report

Once the partitions were up and anchored in place (see drawings accompanying February 15, 1977, Interim Report), we began immediately to install the electrical wiring, boxes, switches, outlets, and fixtures. First, the boxes were attached in the places marked on the studs. Holes were then drilled through the ceiling joists for the wiring. The electrical wiring was strung through the ceiling and down the walls. The fixtures, switches, and outlets were then connected to the wiring in the proper places.

Figure 13.6 depicts the wiring circuit now in the workshop. It was prepared before the electrical work was started. Although the locations of the fuse box, outlets, switches, and fixtures are different in the den, recreation room, and laundry-sewing room, their specifications are otherwise the same.

***Figure** 13.6*
Workshop wiring diagram.

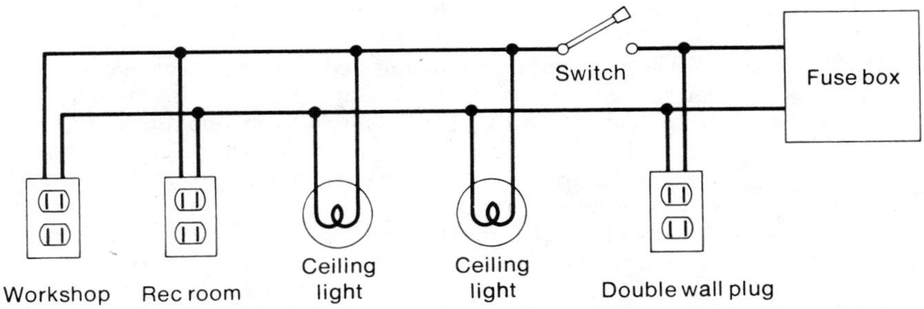

The installation of the wallboard was accomplished without any real problems. Each sheet was measured and cut to fit. The only area in which we had minor difficulties was around the heating ducts.

With the wallboard up, the next stage required much careful, painstaking work. This consisted of taping the wallboard seams and plastering in preparation for painting. Even though these seams were even, great care had to be taken when plastering so that the wall would not have uneven areas. This was done by first spreading a thin coat of plaster over the seams, then pressing the tape into the plaster. When all the air was pressed out from under the tape, and it was even and flat, the tape was covered with another layer of plaster. This layer was sanded down smooth when it dried.

When the plastering was done, we let it dry out; then we started with the painting. It was necessary to apply primer first. By priming, we avoided having to put two coats of paint on the walls. Because the primer paint is less expensive than finish paint, we saved some money doing this.

Problems related to the completion

The main problem encountered since the last report was in stringing the wiring. The 12-2 wire was hard to work with because of its rigidity. It was stiff and hard to bend to thread through the holes in the joists and rafters, especially where close turns were required. Also, we had some problem in installing the fuse box to the existing cement wall. We discovered that the standard screws and wall anchors would not hold the small but heavy box. This problem was solved by first attaching a piece of plywood to the wall and anchoring the fuse box to the board.

The preliminary planning and preparation for this basement remodeling project was the most critical. It was on this phase that the whole project depended for its success. The purpose of each room and the extent of its need was predetermined during this stage. This influenced decisions about the location, size, and furnishings of all of them. The plans used in doing the actual remodeling were based on these decisions; consequently, the kinds and quality of building materials and furnishings were also.

Summary of the main problems, their solution, and how to avoid them in the future

The few problems encountered in doing the remodeling can be considered unavoidable. Perhaps, with more care more flexible wiring could be purchased to avoid the trouble we might have in stringing wire on future jobs. Adjusting walls to compensate for floor unevenness is something usually encountered on every remodeling project. This is true also about warped boards. Corrections of these problems are discussed in our preliminary progress report, dated January 7, 1977.

The den, recreation room, laundry-sewing room, and workshop are now a reality. The small storage room mentioned in the preliminary report and shown on the drawing with it is also completed, with built-in shelving. One or two small pieces of furniture still need to be acquired, but we haven't made up our minds about them as yet. Otherwise, our basement remodeling project was completed on schedule, eleven weeks after its beginning, and all of us, especially our teen-agers, are finding everything quite comfortable.

Conclusion

Application 13–1

Identify each of the following you would develop into a short status report by printing A before it. Place a B before each you would develop into a progress report.

_____ 1. A report of a sales representative on his second day out, identifying the purchasing agents upon whom he called, the products that each ordered, the comments that any may have made about products previously purchased, and where he intends to go on the following day.

_____ 2. A social worker's report about a certain family, identifying its members and describing any physical, social, or economic problems in which any of them may be involved.

_____ 3. A report that presents a description of the work done in an investigation aimed at eventually discovering practical uses of nuclear energy.

_____ 4. A report describing the economic and social conditions of rural Michigan. This report includes a description of the population, income and employment, health, education, and housing.

_____ 5. An insurance report accurately describing the extent of fire damage to a three-bedroom bungalow.

_____ 6. A letter or memorandum from a mechanical engineer in the research department, explaining what he has achieved so far in his search for a more effective design for an automobile muffler.

_____ 7. A brief report that presents the organizational structure of a small hospital, including the professional staff, technical help, and key administrators. The report also includes the main departments and a brief discussion of the duties and responsibilities of each.

_____ 8. The final report of a series describing the last projects that had to be completed before the plumbing installed in a five-story office building would pass state inspection.

_____ 9. A description of a man for whom the Federal Bureau of Missing Persons is searching.

_____ 10. A report giving detailed information about preparations made to begin painting a house.

Application 13–2 Select one of the following and write an appropriate status report. Insert the correct headings and margin notations to identify its main sections and subsections; use the illustrations previously shown in this chapter as guides. Before writing the report, be sure to review the status report format discussed in this chapter. In your report, feel free to use tables, graphs, charts, diagrams, drawings, etc. when needed or helpful.

A. Pretend that you are the chief mechanic of a large trucking company in your city. Write a status report describing the condition of twelve tractor-trailer units. Indicate the number of tractors that are in operating condition and the number that are undergoing repair. Write a brief description of the kinds of repairs being made on the tractors undergoing maintenance. Also, describe the general condition of each trailer.

B. Imagine that you are a nurse or a buyer for a supermarket or department store. Without giving the steps in chronological order, give a descriptive summary of your duties and responsibilities. Indicate the people you come in contact with and the tools, apparatus, or materials you use. Describe what your goals are. If necessary, describe the room, station, office, or laboratory in which you work. Be sure to identify the company and its address. Also, identify the person to whom you are responsible. Indicate the number of people who work under you. Perhaps you can give some statistics as to the number of people you serve or the quantities of merchandise you produce, distribute, or sell during a period of time: a day, a week, a month, or a year.

Application 13–3 Select one of the following and write a status report for it:

A. Pretend that you own or manage a short-order restaurant. Write a general status report that reflects the quantity and quality of food supplies on hand for the coming week. Describe the kinds of canned, fresh, and frozen foods in stock and what you intend to do with them. You might also include foods such as rice, beans, spaghetti, etc., if they are a part of your usual menu. Be sure to define the specific kind of short-order establishment you operate and the kind of customers you mainly aim at satisfying.

B. Imagine that you own or operate a gas station and prepare a status report on the overall business. Indicate from where most of your customers come, the volume of gasoline, oil, and other products you sell. Indicate the number of grease jobs, tire changes, and other services you perform. Discuss how your station compares with others in your neighborhood. You might also describe the building and the main equipment it contains, especially the pumps.

C. Pretend that you are a police officer sent out to report upon an accident. Do not give a chronological sequence of the happenings that caused the accident. Mainly describe the circumstances you found at the scene of the accident upon your arrival, not what was there before or after. Describe the road conditions, the weather, the markings on the road. Give an exact description of the kinds of cars involved and a detailed description of the damage. Give information about the location of the accident, rural, urban, number of intersections, names of streets, and the amount of traffic. Describe the condition of the people involved in the accident. Also, include the names and addresses of witnesses. Be sure not to make a judgment as to responsibility for the accident; that is done in another kind of report to be studied later.

D. Imagine that you are a claim clerk for a motor freight company, and you are to write a damage report on the status of a shipment of living room furniture. Indicate the day and time of inspection, the date the shipment was made, and the carriers who hauled it. Describe how the furniture was packed and the condition of the packaging when you inspected the shipment. Describe the damages to the furniture in detail. Be sure to identify the shipper and the receiver and the freight bill number.

Application 13–4 Study the components of a drafting pencil in Figures 13.7 and 13.8 and write a status report of it. You may treat it as a new pencil or one that has been in use. If you wish, you may trace the parts in the drawing or you may make a drawing of your own for the report.

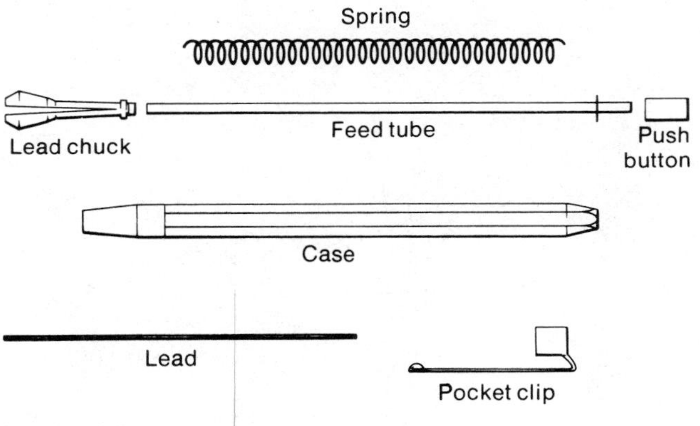

Figure 13.7
Drafting pencil components.

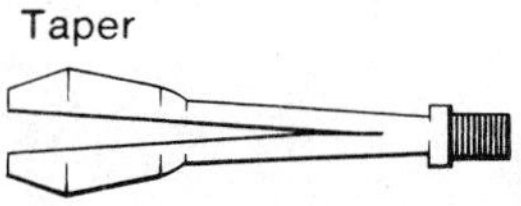

Figure 13.8
Lead chuck for drafting pencil.

Application 13–5 Review the discussions of the various kinds of progress reports in this chapter, especially the suggested formats. After selecting one of the following, write three progress reports for it: a preliminary progress report, an interim progress report, and a final progress report. As you are writing each report, insert the correct headings and marginal notations to identify its main sections and subsections. For any of these progress reports, feel free to use tables, graphs, charts, diagrams, drawings, etc. that you consider needed or helpful.

A. Write three progress reports of a professional house cleaner hired to do someone's spring cleaning. Following are some of the things you might include in them: getting cleaning supplies and equipment, various preparations made before starting, certain precautions taken, a brief summary of the extent of the overall job intended, the amount of time set aside for the job, what was achieved at some interval or interim, problems encountered and how they were solved, the overall cost, and a brief description of the total accomplishments when finished, etc.

B. Imagine that you are a farmer (or a person involved in some other occupation) and write three progress reports on your spring plowing. Following are some of the things you might include in the reports: the size of the area, what you intend to plant in each section, preparations made to get the equipment ready, the quantities and qualities of seeds, fertilizers, and other needed supplies, what you accomplished at some interim period, the problems encountered and how they were solved, what was finally achieved after completion of the spring plowing, the relationship of the kinds of crops planted to future marketing, etc.

Application 13–6 Study Figures 13.9 and 13.10, showing an interior view and floor plan of a barn, and write a complete report discussing its status. Identify each section of your report in the right margin.

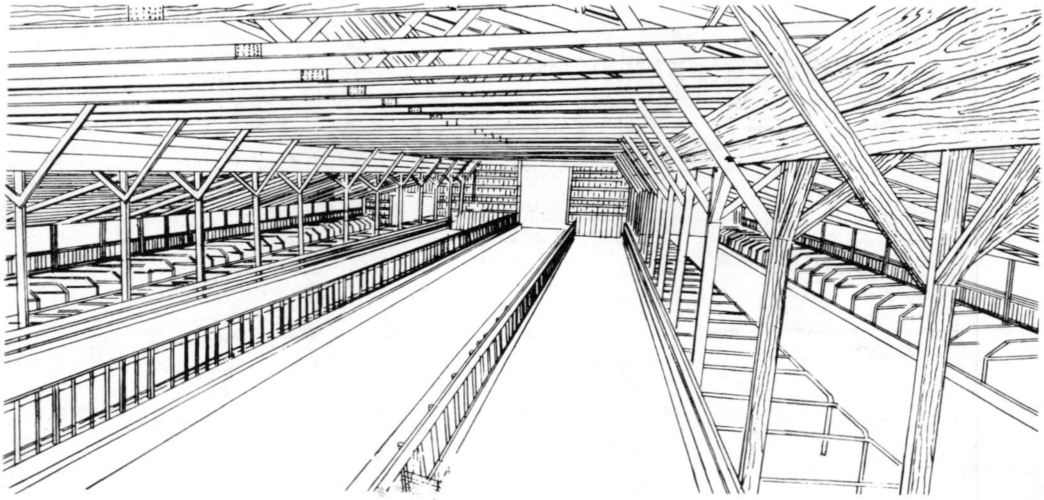

***Figure** 13.9*
Interior view of barn.

SOURCE: U.S. Department of Agriculture, Miscellaneous Publication No. 1241 (Washington, D.C.: U.S. Government Printing Office, September 1972).

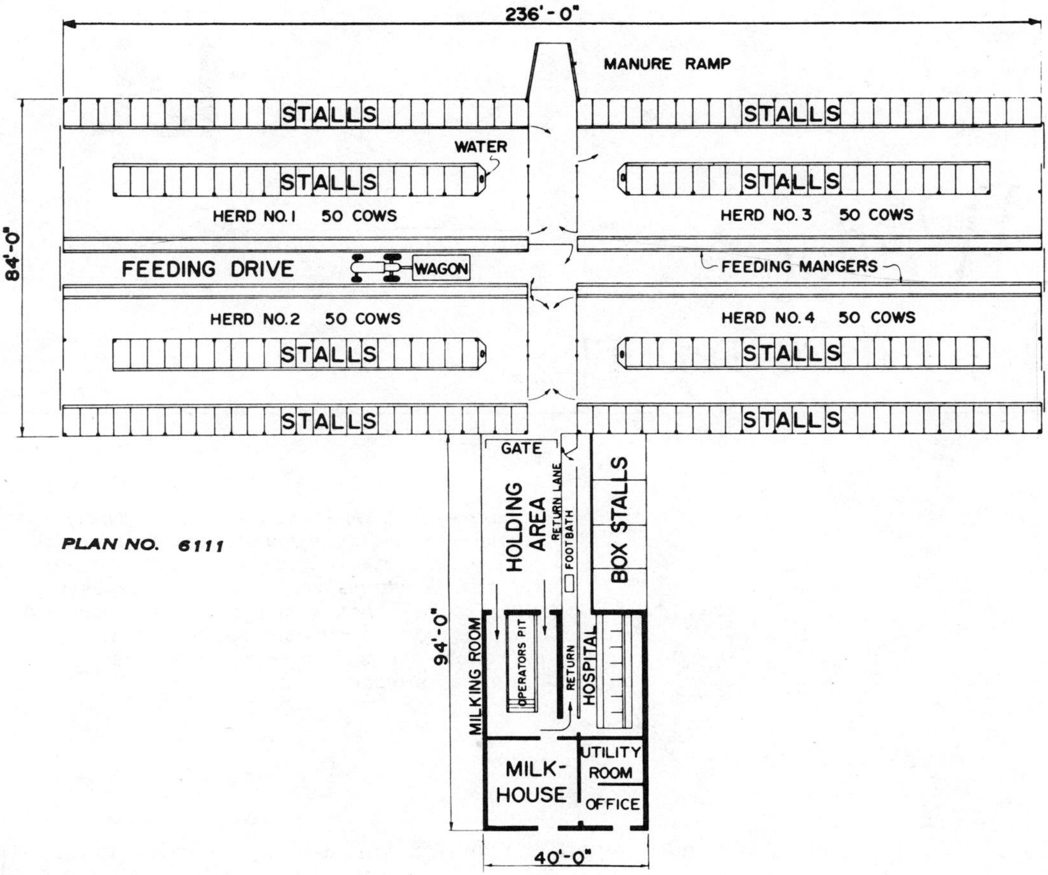

Figure 13.10
Barn floor plan.

SOURCE: U.S. Department of Agriculture, Miscellaneous Publication No. 1241 (Washington, D.C.: U.S. Government Printing Office, September 1972).

14

Writing narrative reports

what they are . . . their purposes . . . kinds: (1) process or procedural and (2) historical . . . kinds of process or procedural: (1) general informative, (2) specific informative, (3) specific directional . . . main sections and organization . . . the functions of each section . . . historical reports . . . format of historical reports . . . example of each kind of narrative report

Just as the status report is mainly a kind of extended description, the reports in this chapter are primarily extended narratives. Like any narrative, they tell in a step-by-step chronological sequence how something begins, develops, happens, or is done. This would be a good time for you to go back and review the discussion of narrative in Chapter 5.

Like status reports, narrative reports are mainly informative reports, not interpretive. They are used to give readers worthwhile information they do not have. The two kinds of narrative reports we will look at in this chapter are process or procedural reports and historical reports. We will carefully examine the writing of three kinds of process or procedural reports: (1) general informative, (2) specific informative, and (3) specific directional.

PROCESS OR PROCEDURAL REPORTS

General informative process or procedural reports are written to readers who do not want or do not need to know exactly how something happens or is done. These reports are much easier to prepare because you won't have to give specific tools, apparatus, materials, or steps in the process. You will know that these details are not needed by the readers; they either can't or don't want to do what you are explaining.

The specific informative process or procedural report is written when the intended reader wants or needs to know precisely how something happens or is done. When one expert tells another how a firefly creates light, he or she may give the exact sequence of occurrences and other exact details explaining the process. When writing one of these reports, you should explain why the main steps occur or must be done, giving the causes and reasons for them.

Specific directional reports, prepared for readers who intend to try doing something, tell them exactly how they should do it. When writing one, therefore, give exact details: tools, apparatus, personnel, material, and ingredients. You may even have to give the specific quantity and quality of each item. To avoid injury or the loss of valuable material, be sure to give the exact steps in the right order and, when necessary, explain the reasons for them. For the same reason, clearly state the precautions to help the reader avoid common pitfalls.

With some minor changes, the following pattern may be used as a guide in your writing of these types of process or procedural reports:

Beginning Section

A. *Title*

Identifies as specifically as necessary the process or procedure to be discussed.

B. *Abstract*

Briefly summarizes the main parts of the process or procedure. Identifies and briefly discusses the final result of the process or procedure, what finally happens or is gained from it. Also, it may concisely restate the important precautions.

C. *Introduction*

1. Identifies as specifically as necessary the process or procedure and what its purpose is.
2. Briefly tells what results from it.
3. Provides the reader with any needed background information: definitions of terms, history, explanation of related principles: laws or theories of chemistry, physics, economics, biology, psychology, business, jurisprudence, etc.

Developing and Supporting Section

1. Identification of the materials and apparatus required by the process or procedure
    a. Identification of the tools, apparatus, and personnel needed.
    b. Identification of the required ingredients, components, and other supplies.

2. Process or procedural narration
   a. A step-by-step narration in orderly time sequence, explaining how the process or procedure takes place or is performed.
   b. If "directional," it gives reasons explaining why certain steps are taken. Necessary precautions should be given emphatically to help avoid possible injury or costly wastage of tools or material.
   c. Helpful graphs, charts, and drawings should be used.

Concluding Section

1. Briefly discusses the results of the process or procedure.
2. Emphasizes the main steps and other important points that should be remembered.
3. States the values the reader can get from the process or procedure.
4. May re-emphasize any important precautions the reader should keep in mind in directive reports.

In the developing and supporting section of the preceding format, you can see the dominant role narration plays in this kind of report.

GENERAL INFORMATIVE PROCESS OR PROCEDURAL REPORT

The following general informative process report explains in general terms the sequence of steps followed in the desalinization of sea water by solar energy. This is not an interpretive report, because the conclusions expressed are a part of the report writer's findings, not those interpreted by the writer from the information found. The conclusions in this report are well-known interpretations made by others. Removing salt from sea water by solar humidification as described in the following report was done by the ancient Phoenicians. This report presents general information; therefore, it is intended for a general reader, not the expert in this field, who already would know all of its contents.

Fresh Water from Salt Water [1]

Beginning Section

ABSTRACT

Today Americans use 360 billion gallons of water a day, three times more than thirty years ago. The present need for water is increasing at a rate of 250,000 gallons per minute. The seas of the world offer the best source for this needed water, and solar energy is one way of desalinizing it. Using solar humidification and solar reservoirs can help solve water-shortage problems today and tomorrow.

[1] Adapted from United States Department of Interior, "The A-B-Seas of Desalting" (Washington, D.C.: U.S. Government Printing Office, 1966).

General informative process or procedural report

Converting salt water to fresh water may be the only way to meet the need for enough water today and the future. With three-fourths of the earth's surface covered by salt water, it is easy to see that countries that touch the oceans have inexhaustible sources for saline water. The United States, for example, has thousands of miles of ocean frontage. All we need to do is to develop a way of removing the salt in a way that will enable us to maintain the costs of production at their present levels. Introduction: purpose statement

Some methods of producing fresh water from salt water have been known for centuries. For example, distillation is one of the oldest ways known for separating fresh water from the saline. In distillation, the water is first boiled and then the steam or water vapor is cooled, condensing into fresh water. Solar humidification was used by the ancient people along the Mediterranean Sea to secure table salt and to run great amounts of desalinized water into their gardens for irrigation. Background information Definition History

Just as solar energy now offers a great source for power, it also may be used for solving our water shortage. The solar humidification process uses the principle that water will evaporate from a surface even when its temperature is below its boiling point. Another principle applied in the desalinization of water is the same as that which keeps a greenhouse warm in cold climates. A greenhouse operates on the principle that the sun does not lose its heat energy to the glass or plastic covering of the structure. The energy of its rays is absorbed by things inside the greenhouse. The darker the interior colors, the more solar energy the objects inside absorb. Principles

The main apparatus needed to convert salt water to fresh water by solar humidification is any device that will cause water to evaporate. Figure 14.1 is a drawing showing two types of solar stills. These are simple devices for the desalinization of water. Supporting Section Materials and apparatus

As the sun's rays radiate through the transparent top of a solar still, the energy is absorbed by a black surface at the bottom. This absorption causes the water resting on the black surface to increase in temperature and to convert Process narration Step-by-step narration

Figure 14.1
Two types of solar stills.

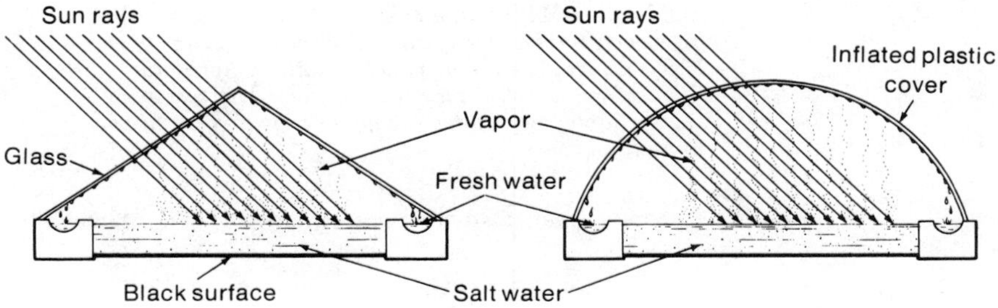

into vapor. This water vapor rises and comes in contact with the top of the still. The still's cover is much cooler than the water vapor. Consequently, when this vapor contacts the cooler top, it condenses and runs down into a collecting trough.

Concluding Section

The advantage of converting salt water to fresh water in a solar still is mainly its low cost. The sun's energy is free. However, the conversion rate depends upon the intensity of the sun's rays. Today, however, experimentation has succeeded in producing cells that are better able to act as reservoirs for surplus solar energy. On days when the sun's rays are not intense enough, the reservoirs of energy may be tapped. Surplus energy from the sun may be stored in these reservoirs for future use. Using solar reservoirs efficiently will enable the solution of a long list of other energy problems along with the shortage of fresh water.

SPECIFIC INFORMATIVE PROCESS OR PROCEDURAL REPORT

When you are asked to write a specific informative process or procedural report, you will be expected to give exact kinds and amounts of materials and apparatus along with the precise sequence of steps. You also will be required to explain why some tools or ingredients are used or why certain steps are taken. This type of report is often written by one expert to help another understand something but not do it.

Even though specific materials and exact steps are given in the next report, it is intended to inform the readers, not to tell them how to make a scuba cylinder themselves. They wouldn't be able to make one of their own even if they wanted to; the costs of the materials and apparatus needed and the amount of work involved would make it almost impossible for a person to make a cylinder of this type. The writer tells the readers why certain ingredients are used and why the steps are done in that order, but the precautions are not emphasized. These precautions need not be stressed because the readers can't or won't want to perform the process or procedure.

Beginning Section

Manufacturing Scuba Cylinders [2]

ABSTRACT

Building a reliable scuba cylinder to withstand 2240 psi of compressed gas requires skillful craftsmanship. It is "self-contained-underwater-breathing apparatus" upon which a skin diver's life depends. The seamless specification necessitates complex, expensive devices to build one.

[2] Adapted from an article appearing in *Skin Diver,* March 1967, p. 46.

Specific informative process or procedural report

Introduction: purpose statement

Though the fancy looking regulator usually gets most of the credit, the scuba cylinder faithfully does most of the work for a diver. It gets slammed about, but continues to do its job faithfully. Little maintenance is required to keep it in working condition. A hydrostatic test every four or five years and an occasional inspection to make sure that it is free of internal rust may be all that is needed to enable it to continue holding a charge of 2240 psi indefinitely.

Background Information

Principle

Some people think that a scuba cylinder was a blob of steel which was later shaped into the form of a cylinder. This is not its history. It is strong and reliable because it is made of seamless, tempered steel. It must be seamless and weldless to avoid any possibility that it might eventually leak under extremely high internal or external pressures.

Developing and Supporting Section

Materials and apparatus (with specific details)

A "why" explanation

A scuba cylinder starts as a 5′ × 10′ sheet of steel plate. This plate is heated to 1300 degrees in an enormous Spheroidized annealing furnace that heats up to 3,000,000 BTU's, and it is kept there for five days. The steel molecules in the sheet are rearranged into a uniform mass of molecular spheres. This annealing process turns the hard steel into soft, workable metal.

Process narration: specific step-by-step sequence

The pliable sheet is allowed to cool; then it is cut into strips. These strips are run through a huge punch press which stamps out blank disks a quarter inch thick and 23¼″ in diameter. Each disk is then conveyed to a hydraulic press weighing 400 tons. It is pressed into a large "cup" shape in this giant machine. Next, it is moved to another press to be "drawn." This is done by a hydraulic ram that pushes the "cup" through a die, narrowing and elongating it.

Another "why" explanation

(Step-by-step continued)

At about this point in the process, the steel "cup" becomes cold. A problem results because the molecular spheres in cold steel get bent out of shape when the metal is pushed around. The cylinder, neckless at this point, has to be put through another annealing process to make it workable again. After this second annealing, the cylinder becomes covered with scale, which has to be cleaned off by dipping the metal into a de-scaling acid. Next it is dipped into a solution to reduce the friction in later draws.

The "final iron" is the last draw in the process of transforming the steel sheet into a perfect neckless cylinder. To form the neck, the steel is shaped by applying direct heat to it. Preheated to a bright red, the lip of the cylinder is placed into a head-forming machine. This machine controls the heat by raising or lowering it as it rolls the lip into a neck.

Next, the cylinder steel is tempered by being placed into a 1650-degree oven. On its removal, it is dipped in an oil bath. Now it passes to other machines that face, drill, and tap the neck. It is then conveyed through a shot blast machine which polishes it off so that it looks like a diving cylinder. Finally, the finished cylinder is given a hydrostatic test under pressure of 3750 psi.

Concluding Section

The reliability of a scuba cylinder results from the way it is made. The next time you see one, you will know that great effort has been exerted to insure its safety. You will know that its durability requires a million dollars in tooling and a tremendous amount of craftsmanship and know-how to manufacture one.

SPECIFIC DIRECTIONAL PROCESS OR PROCEDURAL REPORT

Specific directional reports go one step beyond informative reports by explaining exactly how to do something. When writing one, therefore, you must take care to describe the right tools, material, and steps exactly and clearly enough for the intended type of reader. To the lay person or the apprentice, the directions are given in an informal, more concrete manner. To the expert, however, technical terminology, formulas, and equations help the reader to understand the theory along with the step-by-step procedure.

Specific directional reports are in some ways like the specific informative ones. When writing one, you specify the exact kind and, when necessary, the exact quality and quantity of the tools, apparatus, ingredients, and other materials. You will be expected also to explain the important reasons why something is done or why a material or tool is used.

These directional reports emphasize one type of information that the informative reports do not require. In addition to important details about the tools, materials, steps, reasons why, you often must emphasize the precautions to be taken to avoid physical injury, costly spoilage of materials, damage to hard-to-get tools, and wastage of time and effort.

You probably give specific directions to people almost every day. When you tell someone exactly how to play a certain stroke in golf, how to cook spaghetti, how to clean a carpet, or even how to pass a test, you are communicating a kind of specific directional report.

Before writing a report of this type, be sure to determine how much the readers already understand. If you underestimate what they know, your directions may be repetitious or they may offend the readers with their simplicity. If you overestimate, the readers may become confused because you didn't think it essential to define certain technical terms or to provide necessary background information.

Following is a directional process or procedural report telling readers

Specific directional process or procedural report

how to cross-pollinate. Notice the emphasized precautions. You can see easily that this report is not intended for the expert.

Cross-Pollination as a Hobby [3]

Beginning Section

ABSTRACT

Amateur gardeners can spend many leisure hours trying to form hybrids by doing their own cross-pollination. It is good as a hobby for both summer and winter because it may be done with indoor or outdoor plants.

Purpose statement

The techniques of hybridization by cross-pollination are easy to learn for leisure gardening. Hybridization is an exciting activity because the cross-pollinators never know exactly what will result from their work. However, one is seldom disappointed. More often the gardener is surprised pleasantly by the colorful plant that emerges.

Background information

Definition

Principle

A kaleidoscope is a constantly changing set of colors. Hybridization is like a kaleidoscope in that it is a way in which two or more plants with different qualities are spliced or blended into a single plant. Like the patterns of color that appear to a person peeping into a kaleidoscope, there are an unlimited number of color combinations in the beautiful plants produced by cross-pollination. This type of pollination is achieved by transferring pollen from one plant to the stigmas of another. To be sure the right pollen will fertilize the ovules of the intended female plant, care must be taken to make certain that the pollen of a plant will not reach its own stigmas.

To help you understand how to do your own cross-pollination, Figure 14.2 enables you to see both the male and female organs of a flower in one blossom. Refer to this sketch to clarify any steps that may confuse you in the directions for doing the cross-pollinating.

Developing and Supporting Section

Following is a list of the apparatus and material you will need to cross-pollinate successfully:

Specific apparatus and materials

tweezers
a small camel hair brush
tags with string
hand lens
several small bags
a pencil

[3] The sketch and the information in this report are from United States Department of Agriculture, *Handbook for the Home*, August E. Rehr, "Gardening Galore—Bring the Outdoors Inside: Terrariums, Bonsai, Hybrids" (Washington, D.C.: U.S. Government Printing Office, 1973), pp. 72–80.

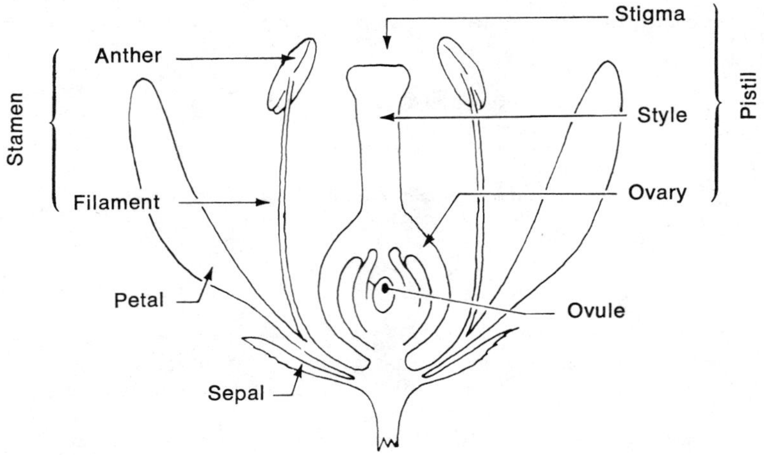

Figure 14.2
The flower shown here is a perfect flower, that is, it has male and female reproductive organs. The stamen is the male organ and the pistil is the female organ.

To make sure that only the proper pollen will fertilize the ovules of the female you select, remove the anthers from the blossom of the female with your tweezers before they shed their pollen. This is especially true if the plant is like the one in the sketch, with both male and female reproductive organs. If you don't do this, you may cause accidentally the cross-pollinization of the same plant. As an added precaution to avoid this, cover the other blossoms with paper or plastic film to prevent pollination by insects or the wind.

Exact step-by-step narration

Precaution

"Why" explanation

Precaution

Next, using your camel hair brush, apply the pollen from the selected male parent to the depollinated blossom of the female parent. This should be done after the surface of the female flower has become sticky, indicating that it is ready to receive the pollen. Now attach a tag to the stem of the plant to identify what you did, the plants involved, and when you did the cross-pollinating.

Step-by-step narration (continued)

A third precaution should be taken to avoid the wrong pollen settling on the parent blossom. Cover the flower again for several days. The fertilization will be completed when the ovary develops. Then, the resulting seeds will be able to produce hybrid plants. If the ovary does not develop, the blossom will just fall off the plant within a few days, and you will have to start the process from the beginning again.

Precaution

Step-by-step narration (continued)

Hybridization is an especially good leisure activity because it can be done throughout the year anywhere. During the cold seasons it can be done indoors, and in warmer

Concluding Section

Specific directional process or procedural report

weather, outdoors. It doesn't require a great deal of equipment, material, physical strength, or money. It is an enjoyable hobby because of the resulting kaleidoscopic patterns of color. If they don't appear the first time you try doing it, keep on cross-pollinating; something will develop. Remember, plants like roses, daylillies, and gladiolus were started this way.

Like all the other suggested patterns of report development discussed so far, you don't have to stick to the preceding or any other format prescribed in this book. Many students and people now in the business world have found these formats to be very practical, and with some modifications constantly use them. Use them until you understand the purpose of each part and of the overall report. Later, you will be able to deliberately and purposefully write and arrange the parts of each report so that they will be most helpful in achieving your purpose.

Short directional process or procedural reports are the ones you probably will need to write most often. Here is another good illustration to help you understand them. The following is a very common kind of directional report. Notice how helpful the drawings are in reducing the amount of explanation in words. They also help readers to understand better the directions verbally expressed.

Building a Compost Tank [4]

Beginning Section

ABSTRACT

Organic mulch is considered to be the best fertilizer by many authorities 'on healthful foods. To avoid using chemical fertilizers in home gardens, build a compost bin in which to allow organic mulch to form. A simple one can be built easily and inexpensively.

Introduction: purpose statement

By building your own compost bin, you can make a convenient device for keeping your mulch in one place as it deteriorates into fertilizer. Using organic mulch in your own vegetable garden is one good way of getting healthful foods. Today, many food and medical authorities are questioning the safety of some chemicals used as fertilizers.

Background information

Definition

Principle

To make your own mulch, the first thing you should have is a compost bin. This is a bin or tank in which to place the leaves, limbs, and twigs from your yard, along with the leftovers from the dinner table while it is decaying and turning into mulch. When it is decayed enough, the mulch is placed around and over plants to nourish them with organic nutrients, needed moisture, and to protect them against the cold.

[4] This report was adapted from "Compost Tank," *Mechanix Illustrated, Workshop Ideas,* Fawcett #723, Fawcett Publications, Inc., Greenwich, Conn., pp. 40–42.

Developing and Supporting Section

Constructing the bin will require only a few hours of work and a couple of simple hand tools. Along with these, you will need to purchase the following:

Materials and apparatus

4 pieces of lumber for legs 2″ × 2″ × 36″
1 panel of exterior plywood ¼″ × 4′ × 8′
16 1¼″ No. 8 screws
waterproof glue
creosote
high grade exterior house paint

The "why" explanation

Plywood is used in building this bin because it is easy to bend. The cross-laminations of the plywood layers are especially suitable for bending it to make the circular design. The plywood should be elevated several inches off the ground. This allows the fresh cuttings and other organic wastes to be dumped on top of the already decaying compost. It also allows the mulch to be removed from the bottom continually.

Specific step-by-step narration

To build this attractive compost bin, take the ¼″ plywood panel and cut it into four pieces. As shown in the Step 1 diagram, the size of each piece should be 24″ × 48″.

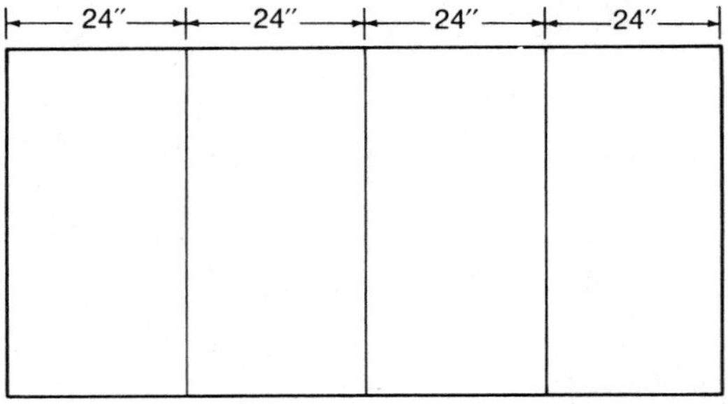

Step 1. Cutting diagram.

Now, as shown in the Step 2 drawing, cut four 2″ × 2″ legs each 36″ long with a 45-degree cut at the top. Next, lap the plywood already cut three inches and connect the legs at each lap with waterproof glue and galvanized screws. Be sure to do this before assembly; it makes the screws easier to drive later.

Precaution

Now lay the pieces flat as shown in the preceding drawing and attach the first three legs. Make certain that the diagonally cut ends of the legs are placed three inches from the top, as shown in the Step 3 drawing.

Precaution

Step-by-Step narration (continued)

Next, bring the end laps of the assembly together and apply waterproof glue to each. Fasten them as shown in the Step 3 drawing and hold them firmly together with two clamps until the adhesive has hardened.

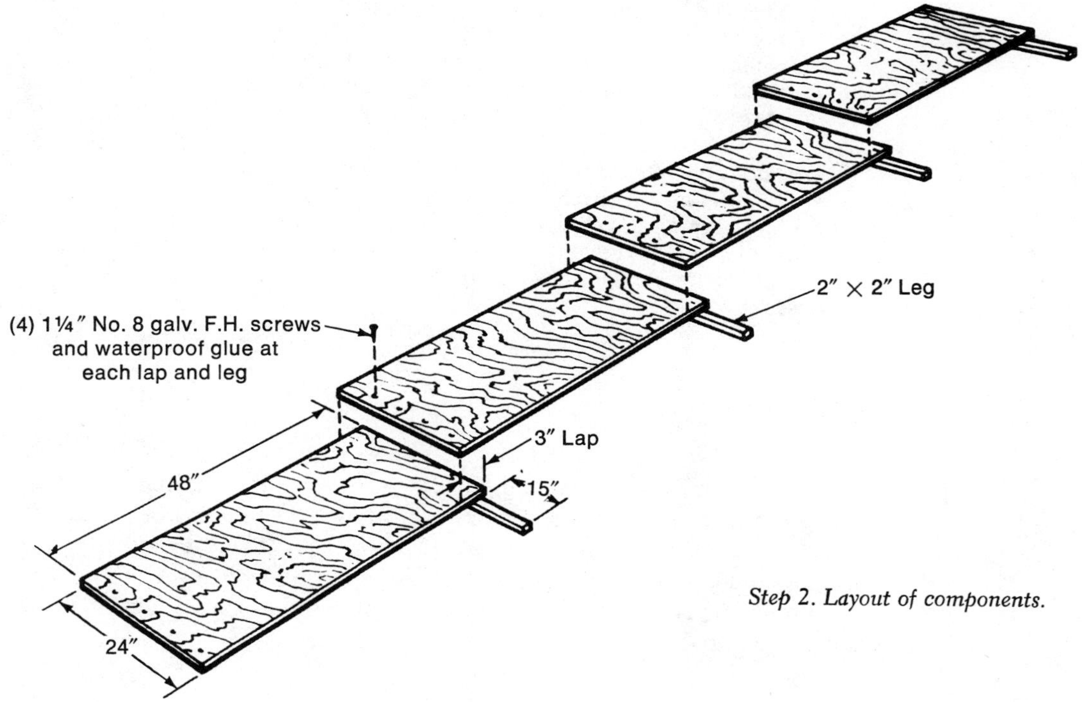

Step 2. Layout of components.

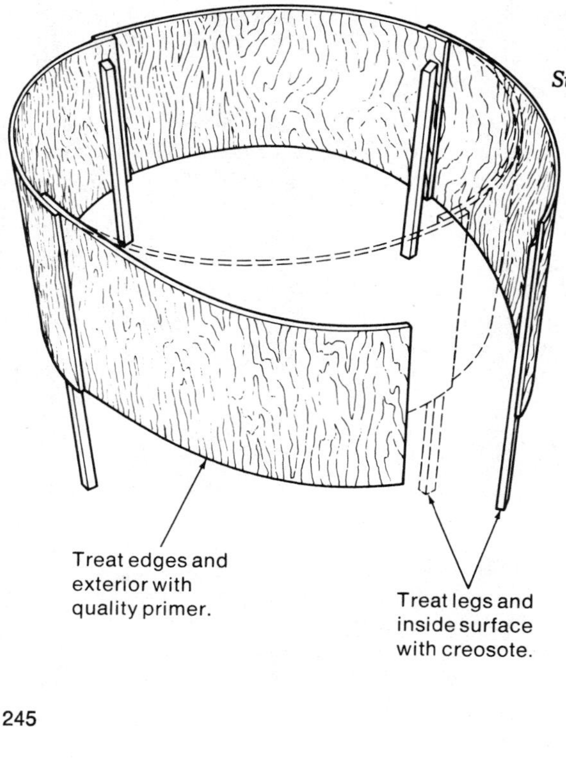

Step 3. Assembly of components.

To protect the legs and inside surfaces of the wood from decay, spread creosote over them. A high-quality exterior primer should be applied to the edges and the exterior surfaces to protect them. For more durable legs, you may use 1¼″ galvanized pipe.

Concluding Section

Sometimes concrete floors are poured for this type of bin, but there is no real need for it. As it is, the bin will do the job of keeping the compost in one place as it slowly decays into fertilizer. It would be a good idea to paint the legs and outside surface with a paint color that matches the house.

HISTORICAL REPORTS

A historical report is somewhat simpler to write than a procedural report. Its main purpose is to trace the origin and growth or development of its subject. It may identify the beginnings of a happening, an invention, a product, a company, or a person. Then it proceeds with a step-by-step narration telling how it changed during a specific time.

Following is the pattern of development you may use in writing short historical reports.

Beginning Section

A. *Title*

Identifies the specific topic and also may give the main change that occurred in it along with the period of time the report covers.

B. *Abstract*

Briefly summarizes only the main stages or incidents in the growth and development of the item discussed. Also, it may describe the final result of the series of happenings or changes.

C. *Introduction*

1. Identifies exactly as necessary what the report is about and the main result of the sequence of happenings or changes.
2. Provides the reader with any needed background information: environment, atmosphere, definition, principles of chemistry, physics, economics, business, etc.

Developing and Supporting Section

A. The Initial Causes or Happenings

Identifies the reasons or causes that started the first incident or happening in the series.

B. Step-by-Step Narration

Traces in chronological sequence the happenings or changes that occurred to the subject. Whenever the interrelationships between the steps in the history are not easily detected, they should be explicitly pointed out.

Concluding Section

A. Briefly summarizes the main happenings and the important changes that took place.

B. May point out how the happenings and changes that have occurred are related to some future event.

C. Restates any main ideas needing additional emphasis.

Like the examples of the other technical reports in this book, the following was deliberately selected because of its shortness. It wouldn't be very practical for you or for the book's publisher to use unnecessarily long reports for purposes of illustration.

Following is a short historical report briefly tracing the development of surveying:

Surveying, Ancient and Modern

Beginning Section

ABSTRACT

Civilization as it exists today could not have developed without accurate surveying and the instruments to achieve it. Orderly interaction of human beings in a restricted area demands recognition of land ownership and boundaries. This need has resulted in the invention and development of the principles and methods for accurate surveying throughout several thousand years of human history.

Introduction: purpose statement

Without the accurate instruments and procedures by which people survey the land upon which they live, today's high degree of civilization would not have been possible. Civilization, the orderly organization and interaction of human beings to promote marked advances in business and science, requires public acknowledgment of land ownership. Before people or their organizations can develop, there must first be accurate definitions of boundaries for ownership and for jurisdiction. The main means by which these land areas are identified is the survey.

Background information

Circumstances

A survey is a graphic representation of an area of land, resulting from determining the position of points on the surface of a particular piece of land. Surveys are used for a variety of purposes, such as the establishment of horizontal and vertical control points, the making of maps and charts, the determination of land areas, and laying out of lines, grades, and detailed dimensions, which serve as guides for construction work.

Definition

Developing and Supporting Section

Initial cause or happening

The need for accurate land surveying instruments was felt early in the history of human existence. The Egyptians as early as 2800 B.C. must have used some type of surveying instrument to accurately build the pyramids. To build them, it was necessary to be able to determine true meridian, and some kind of surveying instruments were needed to do that.

Step-by-step narration

Sometime around 130 B.C. the Greeks developed the *dioptra*, an instrument which could be used either as a level or for laying right angles. The Romans developed numerous surveying instruments to enable them to build roads, aqueducts, and elaborate buildings. Roman surveyors used the *groma*, a useful aligning instrument.

Although there is little evidence to show that progress was made during the Middle Ages in surveying, the following centuries were marked by significant progress. First used in navigation, the magnetic compass was used in the 13th century as a surveying device. The early form of the theodolite, which enables the simultaneous measurement of horizontal and vertical angles, was invented between 1512 and 1572. In 1590, the prototype for the modern planetable was introduced, and in 1690 Galileo built his refracting telescope, which was later used as a land surveying instrument. In 1631, Pierre Vernier invented the vernier, a graduated instrument that is used to indicate fractional parts of divisions, as in a micrometer. Finally, in 1735 the first marine chronometer for the determination of longitude was invented.

Concluding Section

All of the important instruments surveyors use today were developed by the end of the 18th century. Although further refinements have been added to them, these same devices are still in use.

The reports discussed and illustrated in this chapter do not always require as much writing effort as do some others. Although it wasn't easy for you to write the assigned narrative reports, you weren't required to draw logical inferences of your own from the factual details you found and gave to your readers. You simply told clearly how something happens, is done, or should be done. You gave facts and logical inferences made by others; you didn't interpret your findings to arrive at your own logical conclusions and recommendations from them. Any conclusions and interpretations you included in your reports were mainly those made by one of your sources—the person, place, or book from which you got your information.

The reports discussed in the next chapter will require you to go beyond your sources and to present your own conclusions. You will be expected to interpret your findings and to arrive at your own value judgments, to identify causes and reasons to explain a situation, and to recommend problem solutions. At first this may be a bit harder for you than writing narrative reports. It would be much more difficult, however, if you didn't know the differences between the informative reports studied in this and the preceding chapter and the interpretive reports to be studied in the next one.

Application 14–1

Select one of the following or a subject approved by your instructor and write a general informative process or procedural report. In the right margin identify each part of the report, as was done for the illustrations in this chapter.

Voodooism
Viking burial
Egyptian mummification
Building aqueducts in ancient Rome
Manufacturing glass, steel, tin, or other substances.

Application 14–2

Take the topic about which you wrote the preceding general informational report and write a specific informational one for it. Be sure to do the necessary research.

Application 14–3

Examine the accompanying drawing of an auto accident scene and write a specific informational accident report, the kind a police officer might have to make. You might include names of people, streets, automobiles, directions, road and weather conditions, specifics about the extent of injury or damage, and addresses and other information about drivers, passengers, and witnesses.

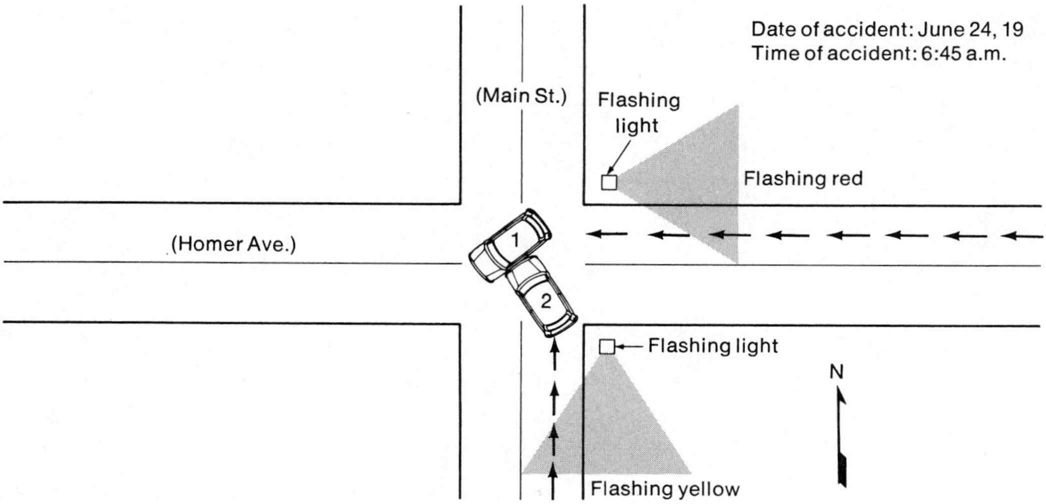

Application 14–4

Review the suggested format for writing a specific directional process or procedural report and the illustration given for one. Next, read the problem below and carefully examine the drawings. Now write a full-length short report of this type telling how the problem can be corrected. Be sure to give the "why" explanations and the precautions. In the right margins of your paper, identify the sections of the report as shown in the illustrations in this chapter.

The Problem:

(1) A lamp or appliance does not work properly.

(2) A broken plug is discovered and must be changed.

Needed Tools and Parts:

(1) Replacement plug.

(2) Knife.

(3) Screw driver.

Procedure:

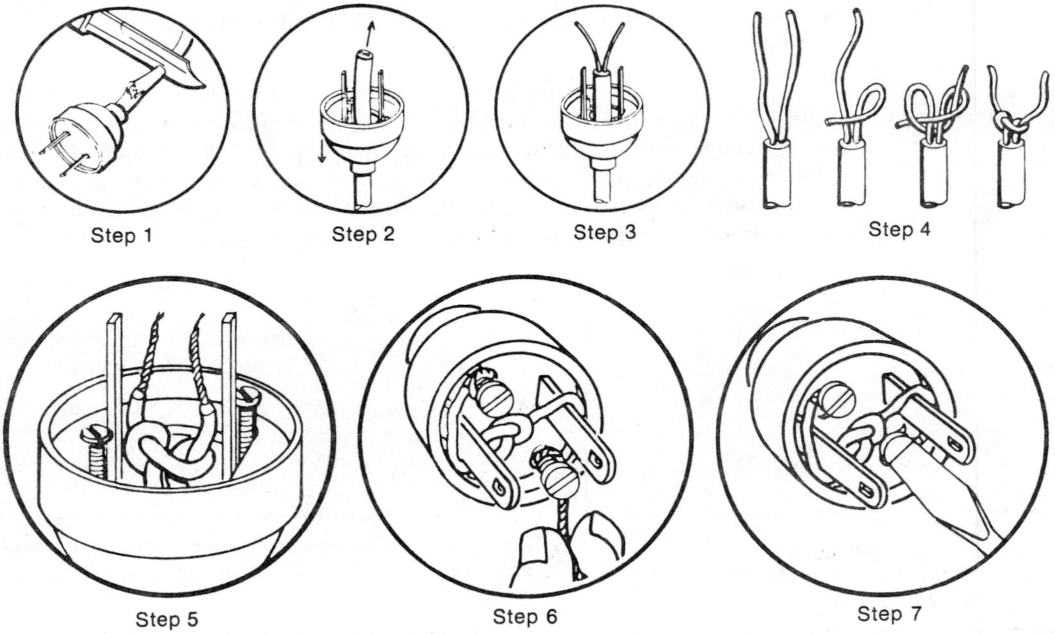

Application 14–5 Write a general informational process or procedural report telling how a student can learn what the school procedures are for securing an on-campus parking permit and what those procedures are.

Application 14–6 Rewrite the preceding report into a specific informational report for a reader who is not a student.

Application 14–7 Rewrite the preceding report again. This time write a specific directional report for new students.

Application 14–8 Write a historical report on the origin and development of a product grown or manufactured in your area, such as corn, apples, soybeans, shoes, tires, or plastics.

15

Writing interpretive reports

what they are . . . kinds: (1) evaluation and (2) problem . . . evaluation report format and main sections . . . kinds of problem reports: (1) causal, (2) proposal, (3) feasibility . . . problem report format and main sections . . . examples of each kind of interpretive report

A technical report writer must know the differences between informative and interpretive reports. In your informative reports, you should present mainly your findings—the facts, inferences, and opinions you detected in your search for information related to your purpose. One or two of the minor conclusions may be your own interpretations, but the important ones will be those you gathered from your sources: the books, persons, places, and objects you examined. An informative report, therefore, is a kind of research paper. Its main purpose is to present what you found in your search for information—to inform, not to persuade. The main purpose of the interpretive report is to persuade readers to accept your conclusions, interpretations, and recommendations derived logically from your findings.

The kind of interpretive report you write will depend upon the main kind of conclusions you want to report. There are two main types of interpretive reports, each presenting its own kind of conclusion: (1) evaluation and (2) problem report. Depending upon its conclusion or recommendation, a problem report may be any of three different types: (1) causal, (2) proposal, or (3) feasibility reports.

It is essential that you be able to differentiate facts, inferences, and opinions to write logical and reliable interpretive reports. Therefore, it would be worthwhile for you to go back at this point and review Chapter 2. Also, since causation plays a vital role in interpretive reports, review what you studied about it in Chapter 5.

EVALUATION REPORTS

The main kind of conclusion presented by an evaluation report is a value judgment. When writing one, you will report your judgment about the usefulness, value, appropriateness, durability, or any of a variety of other qualities of the item you are judging. The main purpose of an evaluation report is to persuade your reader to accept your evaluation.

Since the conclusion, the value judgment, is the main idea in an evaluation report, it should be arrived at logically from sound evidence. Much of this evidence results from applying certain clearly identified standards to the subject of the evaluation. The degree of goodness, fairness, or poorness forming your conclusion depends upon how well the subject lives up to the standards you applied to it. This is like using a ruler, thermometer, or scale to determine the merits of something to persuade your readers your judgment is worth accepting and applying in making decisions about the item evaluated.

When writing an evaluation, you may want to do more than just give your value judgment. You may want to go one step beyond and present your recommendation that your readers do something. You may urge them to purchase or not purchase an item, to adopt or not adopt a policy.

One common mistake in writing evaluation reports is arriving at a subjective conclusion before you identify and apply the standards. You get carried away by emotions. When this happens, the evidence in support of your conclusion is not as reliable as it should be, because it may not be logical. Whether or not you make a recommendation, your value judgment must be supported by reliable evidence. Much of this evidence should result from your logical interpretations of what happens when you measure the degree to which the subject lives up to sound standards, the basis for your judgment.

Most executives and other administrators in business and industry are expected to know how to write a reliable evaluation report. It is a type often used, because raw materials, components, and other products are constantly being evaluated for possible purchase. Also, personnel are often evaluated for possible employment and for promotion. Learn to write good evaluation reports.

As you can see from the suggested pattern for developing an evaluation report below, you should clearly tell not only your value judgment, but also how you arrived at it logically. Especially notice the restatement of the judgment in the concluding section.

Beginning Section

A. *Title*

1. Specifically identifies what is being evaluated.
2. Clearly states the value judgment and, if one is needed, the recommendation.

B. *Abstract*

Identifies the specific subject of the evaluation, the conclusion, and any recommendation. Gives the standards and supports the conclusion by briefly summarizing the extent the subject measures up to the specified standards.

C. *Introduction*

1. Identifies specifically the item being evaluated and the standards by which it is judged.
2. States the value judgment arrived at and if asked for, any recommendation resulting from the evaluation.
3. Provides background information: needed definitions of the item, standards, conclusion or recommendation, historical background, the origin and development of the item or the standards, etc.

Developing and Supporting Section

A. *Justification of Standards*

Justifies the standards or criteria, explains why they are fair, and defends their use, if necessary. This section is used only when the writer senses disagreement with the standards.

B. *Application of Standards*

The writer applies the standards or criteria to the important parts of the item being judged or to the whole item. This is to provide the reader with evidence in support of the judgment, showing why that judgment is correct or valid. In this section the writer may also point out where he or she agrees with those who disagree. Mainly and most important, however, the writer uses this section to provide the reader with enough evidence to support the value judgment and any recommendation given.

The Concluding Section

A. Contains a restatement of the conclusion or evaluation.
B. Includes a summary of the main points in support of the judgment.
C. Submits recommendations that may be required.

Following is a type of personnel report any supervisor might have to write to present an evaluation of a subordinate.

Employee Evaluation Report—John Weston Beginning Section

ABSTRACT

John Weston should be selected for promotion to a foreman position in the Electronics Department. During the years he has worked under my supervision, his knowledge and skills in inspection and quality control of the transistors produced at Tomp-

kins Electronics have been superior. He is especially qualified for this position by communication skills and leadership ability.

John Weston possesses both the technical skill and the leadership ability required by a supervisory position at Tompkins Electronics. I recommend that he be promoted to foreman in the Transistor Department. John is skilled in the manufacture of electrical components and is thoroughly aware of the many applications to which our transistors are put by other manufacturers who use them in their products.

Introduction
Purpose statement
Standards

Conclusion

Background information
Definition

John Weston has been an employee of Tompkins Electronics for the past twenty years. He was hired as an apprentice in April, 1955. Because of his skill and reliability over that period, he has worked his way up through various positions and now is an inspector for the Transistor Department.

History

Any supervisory position at Tompkins Electronics requires that a person be equal or superior to subordinates. A foreman in the transistor department should know about the technology of those transistors so that others can rely on his judgments respectfully. Supervisors should be aware of the principles of human dynamics, enabling one to lead rather than to drive subordinates. These principles and their application will increase wholehearted cooperation and the reduction of the labor disputes which afflict most industries.

Developing and Supporting Section

Justification of standards

It is true that others who have been with Tompkins Electronics for longer periods of time should be given consideration first. Tompkins has always done this, and, as a matter of fact, union regulations demand it. Weston would be the first to urge that this be done. Everyone, however, agrees that the person possessing the best qualifications should be appointed to the position of foreman of the Transistor Department.

Application of the standards

During the years of his employment at Tompkins Electronics, John Weston has contributed greatly to the quality of the electronic components manufactured in his department. After work, he attended classes in electronics at the University of Michigan. This additional training enabled him to contribute significantly to the production methods now used here. These methods have improved the quality of the products manufactured by this company. Seldom have the transistors manufactured by his department been rejected.

As an inspector for the Transistor Department, John Weston is required to submit reports related to production. The many reports he has prepared have been thoughtfully researched and clearly written. They have enabled his superiors and other administrators to make important decisions with confidence, based on his findings, conclusions, and recommendations.

Weston's present position requires that he have more than technical knowledge and skill in writing reports. It also requires leadership ability. He is respected by the people whose work he critically analyzes.

Patiently and carefully he trains new employees. They sense his earnest fairness right from the beginning. Because of his skill, they are not reluctant to seek his advice about the complex problems related to their work.

John Weston should be promoted to foreman in the Transistor Department because he has the best qualifications of anyone eligible for the position. He has both the necessary technical knowledge and skill along with the leadership ability. He has other good qualities that add to making him very qualified for this particular supervisory position.

Concluding Section (with recommendation)

The conclusion in the following evaluation report is a negative evaluation, but it does not make a recommendation either for or against the purchase of the tester. The preceding evaluation report recommends the promotion of John Weston.

A Compression Tester Must Be Durable

Beginning Section

ABSTRACT

Because of the way the check valve is friction fitted to a rubber nipple, the A.C.E. Model 535 compression tester is not durable enough for a busy automobile garage.

The A.C.E. Model 535 compression tester is a poor investment because it lacks durability. Even though it is not built well, it is a popular tool. Much of its popularity is more because of its low cost than its sturdiness.

Introduction
Purpose statement
Topic
Standards
Conclusion

A compression tester is essential for measuring the pressure within an automobile cylinder. Because it is one of the most often used tools, it must be able to withstand hard use in a busy garage.

Background information

Definition

This tester consists of a pressure gauge attached to a tapered rubber nipple with a friction-fitted check valve similar to one on an automobile tire. To check the compression of an engine, the mechanic holds the tester's nipple into the spark plug hole of a cylinder while someone turns the engine over. The gauge, if it's working right, will indicate the highest pressure reached in that cylinder.

Developing and Supporting Section Justification of standards

Durability is very important for any tool used for automotive maintenance and repair. Cost must be given consideration because of the large inventory of tools and other apparatus that is kept available at all times. At today's high wages for a mechanic, a tool becomes very expensive if someone has to run out to purchase a replacement. Replacements for small tools are usually available most of the time, but extra compression testers are seldom kept in stock. Another thing to remember is a broken compression tester becomes even more expensive if it breaks during a test cycle, because it may cause serious damage to the engine being checked.

Some garage and gas station owners argue that they can afford to replace the A.C.E. 535 tester more often because of its low price. Everyone will agree with them on that point. It would still be worthwhile for them to check to be sure that two cheaper testers last at least as long as one of the more expensive ones.

Application of standards

I purchased an A.C.E. 535 tester to check the compression on my motorcycle. While I was checking the pressure on a cylinder, the check valve was forced out by pressure, and it flew into the combustion chamber of the engine. The piston smashed the tester valve into pieces. Since compression checks are done on a warm engine, the heat from the engine can soften the rubber nipple enough to allow this check value to pop out. Serious damage to the cylinder head would have resulted if I had not removed immediately the metal fragments of the smashed valve from the cylinder.

Concluding Section (without recommendation)

Because of this weak connection between the check valve and the rubber nipple, the Model 535 A.C.E. compression tester is poorly designed. It can cause a much more expensive repair job if careful attention is not paid as it is being used.

PROBLEM REPORTS

We have noted that there are three main kinds of problem reports: (1) causal, (2) proposal, and (3) feasibility. The word "problem" here means any question or situation needing investigation. It doesn't always mean something bad, such as the problem of air pollution or that of cancer.

Since the three types of problem reports differ mainly in the kinds of conclusions they contain, all three may follow much the same pattern of development. Following is a format you will find helpful in writing any of these three types of reports.

Beginning Section

A. *Title*

1. Identifies the specific problem.
2. May state the main conclusions or recommendations.
3. For feasibility reports, may identify related time period.

B. *Abstract*

Briefly identifies the specific problem. Summarizes the materials and methods used in conducting the investigation. Briefly states your interpretations in the form of causal explanations or recommendations for proposals and feasibility reports.

C. *Introduction*

1. Specifically identifies the problem, your conclusion, and/or your recommended solution.
2. Provides background information in the form of definitions of the problem, the causes and your other conclusions, or the recommendations, if any are needed.
3. Gives any needed explanations of the principles of chemistry, physics, economics, law, etc. that are related to the problem.
4. Presents any required historical information for the problem.

Developing and Supporting Section

A. The Materials and Methods for Conducting Investigations

1. Tools, Apparatus, Material, and Personnel
   Identifies specifically in sentence or list form the tools, apparatus, and personnel involved in the investigation. It also indicates the quality and quantity of the ingredients, components, and other materials that are a part of the investigation.
2. Methods and Sources
   Identifies the method used in the investigation along with the sources from which the data required for the investigation were derived. It may also tell where the investigation was conducted. This, of course, is in addition to the information in any footnotes or bibliography.

B. Investigative Procedure
Gives a detailed step-by-step chronological sequence explaining how the investigation was conducted. In other words, it is an extended discussion of the procedure or method used to conduct the investigation.

C. Findings and Results
Conveys what was discovered and what resulted from the investigation.

D. Conclusions Based on Findings
(In all problem reports, especially causal)
Identifies and defines the causes for the investigated problem that were interpreted from evidence presented by the findings.

(In proposal reports)
Presents one or more specific recommendations for the solution of the existing problem based on its causes. It may also present values to be gained in support of the recommended solution.

(In feasibility reports)
Presents one or more specific recommendations for the solution of an anticipated problem, one that doesn't exist yet. It also may give reasons supporting the effectiveness and practicability of the recommended solutions.

Concluding Section

A. Summarizes the problem and the main findings of the investigation.
B. Restates briefly the main causes.
C. Briefly urges acceptance of the recommendations in the proposal or feasibility report.
D. May also discuss the relationship of the conclusions or recommendations to the future.

CAUSAL REPORTS

A causal report presents the method and results of an investigation seeking the cause of something. It might report on the way research was done to determine the causes for bird migration. It doesn't have to recommend problem solutions, because its purpose may not be to correct a problem but just to give reasons or causes.

At times you may be asked to conduct an investigation to find the causes for a problem. When this happens, you will be expected to report mainly how you conducted the investigation and the causes you discovered. These causes often will be inferred from the facts and other reliable evidence you found. Your conclusions will depend upon your interpretation of the evidence.

In the causal report you will not be expected to present recommendations for the best solution for the problem. Your main purpose will be to present your findings and conclusions. It will be the responsibility of the person or organization receiving the report to interpret your conclusions, identify various solutions for the problem, evaluate the solutions, and select the best one for use.

Since the causal report presents conclusions derived logically from the findings of your investigation, it is different from the informative report. Evaluation and causal reports are interpretive because they go beyond the findings. The evaluation report makes a value judgment based on the findings, and the causal report draws conclusions from them about the causes for a problem or situation. The causal report, however, does not make recommendations for problem solution.

The writers of the following short causal report identify the problem and present their findings. Notice how the linear graphs clearly present the causes discovered in the investigation. The numbers in parentheses refer to the bibliography listed as "Literature Cited" at the end of the report. They illustrate one type of "footnote" reference discussed in Chapter 12, p. 210.

Causal reports

Biphenyl Absorption of Citrus Fruits: The Effect of Variety, Color, Class, and Injury by Freezing, Peeling, and Lack of Oxygen [1]

Beginning Section

Title

ABSTRACT

Some biological and environmental factors affecting biphenyl absorption were investigated to determine to what extent biphenyl residues in citrus could be controlled during storage and shipping.

Valencia and navel oranges held 14 days in a near-saturated biphenyl atmosphere at room temperature (76° to 84°F) absorbed comparable amounts of biphenyl. Valencia oranges held 6 weeks in a near-saturated biphenyl atmosphere at room temperature absorbed nearly 375 parts per million (p.p.m.) biphenyl. Eureka lemons held under the same conditions absorbed less than 70 p.p.m. biphenyl.

Green oranges and lemons absorbed more biphenyl than the fully colored fruit of comparable size.

Oranges absorbed 3 times as much biphenyl as lemons when held 3 or 4 days in a biphenyl atmosphere. Injury by freezing increased biphenyl absorption by oranges 2 to 4 times and by lemons 6 to 12 times compared with uninjured fruit. Different kinds of citrus absorbed similar amounts of biphenyl when injured by freezing.

Introduction

Purpose statement

The problem

Biphenyl has been used as a fungistatic agent in the control of citrus fruit decay since Tomkins first reported its effectiveness in 1935 (11). In commercial practice, pads of kraft paper treated with biphenyl are placed inside cartons of citrus fruit during packaging. The biphenyl volatizes from the pads into the atmosphere of the carton. The biphenyl vapor controls decay in citrus fruit through inhibition of fungal vegetative growth and normal spore formation (7). At the same time, however, biphenyl is absorbed by the citrus fruit in the carton. The accumulation of biphenyl in the fruit must be kept below the legal tolerance of 110 p.p.m. for the United States and 70 p.p.m. for the European Economic Community (EEC). Previous research (9) has indicated that oranges might exceed the EEC's tolerance because of the prolonged storage required for overseas shipments.

History (other investigations)

The amount of biphenyl absorbed by the fruit varies with storage conditions and kind of citrus. Rygg and others (10) reported that oranges absorb 2 to 5 times as much biphenyl as lemons and twice as much as grapefruit under the same storage conditions. Hayward and Edwards (3) compared biphenyl absorption by 14 different varieties and

---

[1] Shirley Norman, C. C. Craft, D. C. Fouse, Agricultural Research Service, U.S. Department of Agriculture (Washington, D.C.: U.S. Government Printing Office, March 1971).

kinds of citrus during 2 weeks' storage at 70°F. Kumquats, tangerines, and possibly tangelos absorbed biphenyl in excess of 110 p.p.m. Valencia oranges absorbed about 70 p.p.m., while Hamlin oranges, Duncan and Marsh grapefruit, and Temple oranges absorbed 40 to 60 p.p.m. biphenyl. Pineapple oranges and limes absorbed 30 to 40 p.p.m. and lemons absorbed less than 20 p.p.m. biphenyl. Hayward and Edwards (4) reported that tangerines absorbed twice as much biphenyl as oranges and 4 times as much as grapefruit under the same conditions.

Definition of the problem

Some biological and environmental factors affecting biphenyl absorption were investigated to determine to what extent biphenyl residues could be controlled during storage and shipping. Tests were undertaken to determine the effect of (1) color class, (2) 2-chloroethane phosphonic acid (Ethrel), (3) oxygen-free atmospheres, and (4) mechanical and freezing injury on biphenyl absorption by citrus fruits. In addition, the daily and weekly accumulation of biphenyl by oranges and lemons was followed in a near-saturated atmosphere of biphenyl to determine how much biphenyl would be absorbed by citrus fruits.

Developing and Supporting Section

Materials and apparatus

The biphenyl chamber consisted of a large enclosed metal cabinet. Biphenyl crystals were spread evenly on the bottom of the chamber. An open container of biphenyl crystals was placed inside at the top of the chamber. Because of the restricted ventilation of the chamber, the biphenyl vapor in the atmosphere inside the chamber was probably near saturation. The maximum concentration of biphenyl vapor possible at room temperature is approximately 73 to 79 micrograms per liter (1). Wire baskets containing the citrus fruit were placed on shelves inside the chamber. The fruit did not come in contact with the biphenyl crystals.

Anaero-Jars, size No. 3, 5-gal. capacity were used in place of the chamber when an atmosphere other than air was desired. Humidified air or nitrogen was flushed through the jars at a rate of 200 milliliters (ml.) per minute.

Methods

Since it was desirable for the fruit to absorb a maximum amount of biphenyl in the chamber and to have a maximum amount of biphenyl vapor available in the jars, room temperature (76° to 84°F), was chosen for all tests except one. Green and fully colored oranges were held at 50°.

Biphenyl pads, each containing 2.38 ± 0.07 grams of biphenyl, were used as a source of biphenyl vapor in the jars. Four pads were placed inside each jar.

Unwaxed fruit, washed by standard packinghouse procedures, were used except for the green and fully colored fruit which were hand picked in the orchard and washed in the laboratory. Green and fully colored fruits were of comparable size.

Biphenyl residues were determined by thinlayer chromatography and spectrophotometric measurements according to Norman and others. Duplicate samples of 300 grams of fruit were analyzed. Biphenyl content represents parts per million of biphenyl on the basis of fresh fruit weight.

The citrus fruits were submerged in a water solution containing 5,000 p.p.m. 2-chloroethane phosphonic acid (Ethrel) for 2 minutes and then air dried.

Procedure of investigation (chronological step-by-step sequence)

The absorption of biphenyl in a near-saturated biphenyl atmosphere was determined for the following treatments:

1. Valencia and navel oranges and Eureka lemons were held in the biphenyl chamber 6 weeks and analyzed daily and weekly.
2. Green and fully colored Valencia oranges and Eureka lemons were compared after 7 days in the biphenyl chamber.
3. Valencia and navel oranges, Lisbon and Eureka lemons, and Satsuma tangerines were frozen at −10°F, thawed overnight, and then held 3 days in the biphenyl chamber along with uninjured control fruit.
4. Valencia and navel oranges and Eureka lemons were held 4 and 7 days in jars flushed with humidified air or nitrogen.
5. Valencia orange peel and pulp, separated carefully by hand, were compared with whole fruit after storage for 3 days in the biphenyl chamber.
6. Valencia and navel oranges and Eureka lemons, untreated and treated with Ethrel, were held 4 and 7 days in jars flushed with humidified air.

Findings and results

Supporting causal evidence

The absorption of biphenyl by Valencia and navel oranges on a daily basis for a period of 14 days at room temperature is shown in Figure 15.1. The rate of absorption was somewhat higher at the beginning than after the fruit had been in the biphenyl atmosphere for a few days. Very little difference in absorption was noted between Valencia and navel oranges. These oranges absorbed about 75 p.p.m. biphenyl by the 7th day and 106 to 134 p.p.m. by the 14th day in the biphenyl atmosphere.

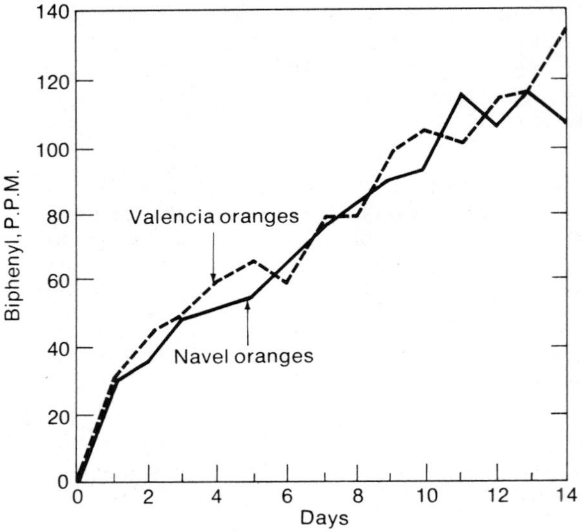

Figure 15.1
The absorption of biphenyl by Valencia and navel oranges held in a near-saturated biphenyl atmosphere 14 days at 78° ± 2°F.

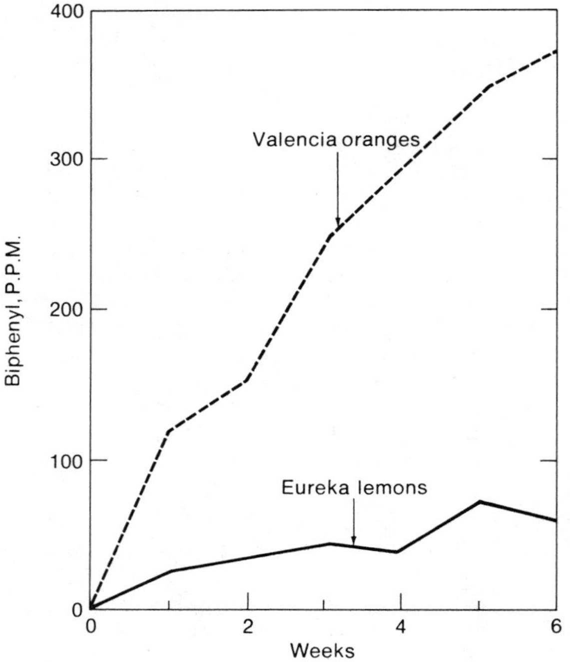

Figure 15.2
The absorption of biphenyl by Valencia oranges and Eureka lemons held in a near-saturated biphenyl atmosphere 6 weeks at 82° ± 2°F.

Conclusions based on findings

The absorption of biphenyl by Valencia oranges and Eureka lemons on a weekly basis for a period of 6 weeks is shown in Figure 15.2. The oranges lost 10 percent moisture in the 6 week period and the lemons, 16 percent. The biphenyl data are corrected to the initial weight basis.

Causal reports

Room temperature was somewhat higher during this test than that shown in Figure 1. The oranges absorbed nearly 120 p.p.m. biphenyl in 1 week and nearly 375 p.p.m. biphenyl by the end of the 6-week period. The lemons approached 70 p.p.m. biphenyl after 5 weeks but never exceeded this amount. The oranges absorbed 4½ times as much biphenyl as the lemons during the first 2 weeks, 5½ times as much the third week, and 6 times as much during the last three weeks of storage.

The absorption of biphenyl by different color classes of Valencia oranges and Eureka lemons was compared after storage for 7 days in a biphenyl atmosphere at room temperature. One lot of green lemons absorbed 150 p.p.m. biphenyl or twice the amount of biphenyl absorbed by fully colored lemons (70 p.p.m.). Another lot of green lemons absorbed 81 p.p.m. biphenyl or 5 times as much as that absorbed by fully colored lemons (16 p.p.m.). The difference in absorption between green and fully colored oranges was not as great as it was with lemons. Green oranges absorbed 55 p.p.m. or 1.4 times as much biphenyl as that absorbed by fully colored oranges (39 p.p.m.) after two weeks storage at 50°F. The absorption of more biphenyl by green lemons and oranges than by fully colored fruit agrees with data reported by Rajzman.

Concluding Section

Summary of findings and conclusions

These experiments indicate that (1) oranges may absorb as much as 385 p.p.m. biphenyl in 6 weeks at room temperature and possibly more with additional storage time, (2) lemons absorb considerably less biphenyl than oranges, (3) green oranges and lemons absorb more biphenyl than fully colored fruit, (4) biphenyl absorption by citrus fruit is increased by mechanical injury, injury by freezing, and injury from lack of oxygen, and (5) increasing the respiration rate of oranges and lemons with Ethrel treatment does not increase biphenyl absorption.

LITERATURE CITED

(1) Bradley, R. S., and Cleasby, T. G.
1953. The vapor pressure and lattice energy of some aromatic ring compounds. Jour. Chem. Soc. Pt. II: 1690–1692

(2) Denny, F. E.
1924. Effect of ethylene upon respiration rate of lemons. Bot. Gas. 77: 322–329

(3) Hayward, F. W., and Edwards, G. J.
1963. The effect of time and temperature of storage on residues of diphenyl in citrus fruits packed in cartons with diphenyl pads. Fla. State Hort. Soc. Proc. 76: 318–320

(4) ——— and Edwards, G. J.
1964. Some factors affecting the level and persistence of diphenyl residues on citrus fruits. Fla. State Hort. Soc. Proc. 77: 315–318

(5) Norman, S., Rygg, G. L., and Wells, A. W.
1966. Improved cleanup method for determination of biphenyl in citrus fruits and in biphenyl-impregnated kraft papers by thin layer chromatography. Assoc. Off. Analyt. Chem. Jour. 49(3): 590–595

(6) Rajzman, Anna
1965. Les residus de biphenyl. Residue Rev. 8: 1–73

(7) Ramsey, C. B., Smith, M. A., and Heiberg, B. D.
1944. Fungistatic action of diphenyl on citrus fruit pathogens. Bot. Gas. 106: 74–83

(8) Russo, L., Dostal, H. C., and Leopold, A. C.
1968. Chemical stimulation of fruit ripening. Bioscience 18: 109

(9) Rygg, G. L.
1969. Biphenyl content, transit temperatures, and fruit condition of California citrus in overseas shipments. U.S. Dept. Agr. Market Res. Rpt. 830, 22 pp.

(10) ———, Wells, A. W., Norman, S. M., and Atrops, E. P.
1964. Biphenyl control of citrus spoilage: Influence of time, temperature, and carton venting. U.S. Dept. Agr. Market Res. Rpt. 646, 22 pp.

(11) Tomkins, R. G.
1935. Wraps for the prevention of rotting fruits. Gt. Brit. Dept. Sci. Ind. Res., Food Invest. Board Rpt.

PROPOSAL REPORTS

A proposal is a kind of problem solution report. It may be an intercompany report, one prepared by someone outside an organization. The report describes a specific problem, tells how the investigation was conducted, identifies the findings and the conclusions based on them, and presents the recommended solution.

Before writing a proposal, you should make a careful examination of the invitation or authorization to study the problem so that you may better understand it. An "on-the-site" inspection of the circumstances along with interviews with employees closely related to the problem would be helpful.

If you cannot arrive at a clear solution recommendation for the problem, it is better to indicate that in your report. In such instances it might be wise to provide the reader with your findings, the conclusions based on them, and the various possible solutions you think worth considering. By doing this, you will be allowing your reader to make the decision as to which is best.

The writer of the following report was called upon to make observations in an accounting department, to investigate a problem, and to write a proposal report with a recommendation for its solution. The writer gives the recommendation and extensively supports it by enumerating the rewards or gains under a "Systems Advantages" heading. Notice how clearly the problem is identified. The person who authorized the report is also mentioned.

Proposal Report for Mechanization of Inter-Intra Company Accounts Payable System

Beginning Section

ABSTRACT

The Detroit General Parts Warehouse has outgrown the usefulness of its present accounting system. The MIT 432 Electro-Calculator will solve many of the resulting problems by requiring fewer pieces of equipment, requiring less personnel, and increasing work speed and accuracy. The cost of conversion to the MIT system can be reduced by retaining the compatible features of the present system so that the change can be made in stages.

Introduction
Purpose statement
The problem
Proposal authorization

To efficiently handle computational and posting functions of inter-intra company accounts payable, one MIT 432 Electro-Calculator with card punch and posting carriage is recommended to replace the Lasher accounting machine now in use. The submission of this report was requested by Mr. A. W. Mather, Panda Motors General Parts Warehouse, Detroit, Michigan.

Developing and Supporting Section

In general, the present procedure for processing invoices received from other Panda Motor divisions is as follows:

Investigation findings:
Procedure
Apparatus
Personnel

1. Upon receipt of goods an MIT card is created showing, among other things, a bates number and a dealer net. It is used for figuring markup after processing of invoices.
2. When invoices are received, a comptometer operator checks each line extension, and a distribution code is assigned where applicable.
3. After authorization for payment, invoices are posted to account ledgers and remittance advice forms (carbonized).
4. While this posting is being done, distributions are made to the various registers of the Lasher accounting machine.
5. At month end, a settlement certificate is prepared on all open items for the month on the posting machine.
6. The difference between dealer net and cost is calculated manually after the posting and distribution.

Personnel time involved

1. The time of 1½ comptometer operators.
2. 24 hours of account posting.

3. Month end settlement certificate preparation time.
4. Markup calculation time.
5. Authorization procedure time.

Proposal (solution recommendation)

With the MIT 432 invoice extension verification, account posting, settlement certificate preparation, and distribution through coded punched cards would be accomplished by one operator. No special training would be required to operate the following pieces of equipment:

1. An eight-core storage calculator, with the capability to add, subtract, multiply, round-off, cross foot, and carry credit balances.
2. A modified MIT Elec-typewriter with special program tape reader.
3. A companion numeric keyboard for entry of non-alphabetic data.
4. A front feed posting carriage for utilization of ledger and other accounting records.
5. And a cable connected printing keypunch, which automatically creates 80 column punched cards as a by-product of postings or calculations.

Following are the steps in the improved procedure:

1. Invoices would be authorized for payment, grouped as determined by control classification, and coded for distribution.
2. The 432 operator would process each invoice completely from verification, to posting to accounts, in the following manner:
   a. On a journal sheet at the left of the posting carriage, each extension would be calculated, and codes entered on the numeric keyboard. Also, distribution codes would be entered, and as a by-product, a punched card would be created that would contain the information necessary for sorting and tabulating by MIT data processing to determine distributions and, after merging with receiving cards, markup calculation.
   b. After extensions were calculated, a total for the invoice would be established to agree with the invoice total.
   c. With this total stored in the calculator, the carriage would tabulate to the ledger posting position, and the invoice amount would be posted to the ledger and remittance advice.
   d. Daily or group control totals would be accumulated in the 432.

e. Extension calculators need not be reverified because each factor prints and could be sight checked.

3. Settlement certificates could either be created as they are now, or through the cards produced they could be done by the MIT Tab Department.
4. An important benefit of the cards created, in addition to the distribution advantages, would be the ease with which costs can be run against dealer nets by the tab department.

Rewards or gain supporting the proposal

A. There would be a definite personnel saving using the MIT 432. Since workloads are now distributed among several employees, it is not possible to attach a dollar savings to the proposed system; however, since the inter-intra company payables workload is being done by several employees, it is obvious that a one-operator system would mean considerable per hour savings.

B. By combining the verification and posting, time would be saved in balancing posting to calculations and add-lists.

C. The punch card code distribution would not only save time in posting, since the operator would not need to mechanically select registers, but greater accuracy of account distribution would be established.

D. By punching costs in the output cards, the time-consuming dealer net-cost calculations would be reduced to a simple MIT tab run.

E. Combining several operations into one would make it easier to establish control and affix responsibility to one individual.

F. The best features of the present system, such as historical account records, would be retained, but the disadvantages of mechanical distribution would be eliminated.

G. Any program changes, brought about by procedural or intercompany procedure change, would be made without cost. This is possible because of the built-in flexibility of the 432 program reader.

H. Because the 432 utilizes the standard electric typewriter and 10-key numeric keyboard, as well as a prepunched program, it is extremely simple to learn and operate. Operator decision is reduced to a minimum.

I. Through the total cards produced from posting, a tab run would show unpaid or open items received during the month.

Concluding Section

Up to the present time, the Lasher accounting system and machines may have served Panda Motors satisfactorily.

However, with the extensive increase in the responsibilities of the Detroit General Parts Warehouse, the Lasher system is no longer adequate. The MIT 432 Electro-Calculator system will not only increase the efficiency of the operation of the General Parts Warehouse, it will do it with less help and few pieces of equipment. It does not require abandonment of the best features of the present system, but it does eliminate the inaccuracies of its mechanical distribution. The time-saving, work-reducing, accurate and flexible features of the MIT 432 system are comparatively simple to learn and operate. This simplicity is one of its best selling points.

FEASIBILITY REPORTS

The third kind of problem report identified earlier in this chapter is the feasibility report. Large businesses or other organizations cannot wait until a problem really exists before solutions for it are found. Feasibility reports are prepared to offer solutions for anticipated problems, those that are expected but that may never arise. These are contingency reports: those offering solutions if certain conditions occur. In a sense, they are like "fire drill" practices. The writer says that if such and such happens sometime in the future, here is what we can do to meet and overcome it.

Feasibility reports are prepared by various organizations to deal with important potential happenings. Some of these occurrences may require immediate construction of buildings, quick purchase of many new machines, hiring of squads of skilled experts, and the securing of rare tools and raw materials. Waiting to make these arrangements until the need actually arises might be tragic, even fatal.

Feasibility reports differ from proposals in two main ways. (1)Feasibility reports are usually intracompany, originating inside the organization. Proposals are most often submitted from the outside. (2) Feasibility reports result from a study to determine whether a problem can be solved, and if it can be, to explain and recommend solution should the problem ever become a reality. Also, the feasibility report explains in detail whether or not a certain solution will work or whether it can be accomplished. A proposal report, however, mainly provides the details of an investigation, along with findings leading to conclusions about the solution of an existing problem. Its solution recommendations are for a real situation.

The following feasibility report proposes a solution for the energy shortage, a situation rapidly becoming a serious problem. There are still possibilities for solving the problem before it becomes critical; nevertheless, plans must be made just in case.

Harnessing Geothermal Energy for Future Power [2]

Beginning Section

ABSTRACT

Because of the rapidly decreasing supply of fuels from which to manufacture electrical energy and the increasing demands for more power, the crisis of energy is already on the world's horizon. Governments and industry are vigorously trying to find new sources. Geothermal energy, more predictable than atomic energy, can become the most abundant kind in many areas of the earth.

Introduction

Purpose statement

Feasible proposal recommended

With increasing population and industrial expansion, domestic requirements for electric power have been doubling about every ten years. To meet these growing needs, government and industry are vigorously investigating and rapidly developing new sources of energy. Among the possible new sources, atomic energy probably has the largest potential, but geothermal energy—a previously little explored source—may prove to be most important in many areas.

Background information

History

For years man has viewed with awe the spectacular bursts of natural steam from volcanoes, geysers, and boiling springs. Although the use of hot springs for baths dates to ancient times, the use of natural steam for the manufacture of electric power did not begin until 1905. That year the first geothermal power station was built at Larderello, Italy. For the next several decades, there were no other major developments in the field, and even now Italy leads the world in power production from natural steam. New Zealand began major exploration of hot spring and geyser areas in 1950, and successful results there proved that commercial steam can be developed from areas containing very hot water rather than steam at depth. Today, the United States, Japan, and the Soviet Union are also producing power from geothermal sources, and Iceland uses hot water from geyser fields for space heating. Many other countries have geothermal energy potential, and several are now conducting exploration for sources to be developed.

In the United States, the first commercial geothermal power plant was built by the Pacific Gas and Electric Co. in 1960 at "The Geysers," California, utilizing geothermal steam purchased from the affiliated Magma Power and Thermal Power Companies. The capacity of the plants in 1967 was 54,000 kilowatts.

---

[2] Adapted from: U.S. Department of the Interior, "Natural Steam for Power" (Washington, D.C.: U.S. Government Printing Office, 1973).

Principle (geological)

In a general way, geothermal fields are either hot-spring systems or deep insulated reservoirs that have little leakage of heated fluids to the surface. Yellowstone National Park and Wairakei, New Zealand, are examples of large hot spring systems. Larderello in Italy and the Salton Sea area of California are examples of insulated reservoirs.

Developing and Supporting Section

Investigation

Findings and results

Mineral exploration over the world has shown that temperatures in deep mines and oil wells usually rise with increasing depth below the surface. One popular explanation assumes that our planet has a fiery origin and that a shallow crustal layer encases a large molten core. Most geologists, however, now believe that our planet was not hot when it first formed. The weight of the evidence suggests instead that a natural radioactivity, present in small amounts in all rocks, has gradually heated the earth, and that heat is still being produced. Geophysical studies also indicate that the molten core is much smaller than was once supposed, and that it is not, in itself, a source of the heat in the earth's crust. The reasons for the existence and specific location of the earth's volcanic belts are still subjects of vigorous scientific study and controversy, but the energy from natural radioactivity in rocks of the earth's crust and upper mantle is the fundamental cause of heat within the earth.

Temperatures in a deep well or mine increase, on the average, by about 1°F for each 100 feet of depth. For example, if a deep well is drilled at a place where the average surface temperature is 50°F, we might expect to find a temperature of 212°F at a depth of about 16,200 feet (an increase of 162°F added to the surface temperature of 50°F).

Great temperature differences occur where water can seep underground and can circulate to great depths. Such water is heated as it travels downward; it may return to the surface to be discharged as a hot spring. If the heat supply is unusually great, as it may be in the vicinity of a large body of molten or recently congealed magma, the water may become so hot that near the surface it erupts as a geyser. Near hot-spring systems, temperatures may increase locally at rates of several degrees for each foot of depth, but these high rates of increase do not extend downward for great distances.

Hot springs have a plumbing system of interconnected channels within rocks. Water from rain or snow seeps underground. If the water reaches a local region of greater heat, it expands and rises, being pushed onward by the pressure from new cold and heavy water that is just entering the system. The hot water is discharged as hot springs or geysers.

Feasibility reports

Causes and reasons supporting the feasibility of the recommended solution

Most of the promising areas for geothermal power development are within belts of volcanic activity. A major belt called "the ring of fire" surrounds the Pacific Ocean. The "hot spots" favorable for geothermal energy are related to volcanic activity in the present and the not-too-distant past. In the western United States, particularly along the Pacific Coast, widespread and intense volcanic activity has occurred during the past 10 million years. The record of volcanism in our western states, therefore, holds promise for geothermal power development. Currently, exploration for power sites is focused in California, Nevada, Oregon, and New Mexico, with some interest being displayed in the whole region from the Rocky Mountain to the Pacific Ocean.

Most known geothermal reservoirs contain hot water rather than steam. Water at depth and under high pressure remains liquid at temperatures far above 212°F, the boiling point of water at sea level. When this water is tapped by drilled wells and rises to the surface, the pressure falls. As the pressure decreases, the water boils, perhaps violently, and the resulting steam is separated from the remaining liquid water. Because the well itself acts as a continuously erupting geyser, the expanding steam propels the liquid water to the surface, and pumping costs are nil.

Because of the pressures at great depths, water can be entirely liquid rather than steam deep in hot springs and insulated reservoir systems, even at very high temperatures. Steam forms in these systems if the hot water rises to levels where the pressure drops to the point where water can boil. This flashing of steam from liquid water is the major potential source of geothermal energy for commercial use, because natural hot water systems are relatively abundant.

However, in a few explored systems the heat supply is so high and the rate of discharge of water is so low that steam forms deep in the system. Larderello in Italy and "The Geysers" in California are examples of the less common reservoirs of dry natural steam.

The most favorable geologic factors for a geothermal reservoir of commercial value include:

1. A potent source of heat, such as a large chamber of molten magma. The chamber should be deep enough to insure adequate pressure and a slow rate of cooling, and yet not too deep for natural circulation of water and effective transfer of heat to the circulating water. Magma chambers of this type are most likely to occur in regions of recent volcanism, such as the Rocky Mountain and Pacific States.

2. Large and porous reservoirs with channels connected to the heat source, near which water can circulate and then be stored in the reservoir. Even in areas of slight rainfall, enough water may percolate underground to sustain the reservoir.
3. Capping rocks of low permeability that inhibit the flow of water and heat to the surface. In very favorable circumstances, cap rocks are not essential for a commercial field. However, a deep and well-insulated reservoir is likely to have much more stored energy than an otherwise similar but shallow and uninsulated reservoir.

Concluding Section

It is too early to judge whether natural steam has the potential to satisfy an important part of the world's requirements for electric power, but in locally favorable areas it is already an attractive source for cheap power. Current exploration, based upon geologic and geophysical methods, is likely to develop presently undiscovered fields. The recent discovery of a new field at Monte Amiata, Italy—where there are only meager surface manifestations of abnormal geothermal energy—was based in part on the use of such methods. These are now well enough developed to support exploration for wholly concealed reservoirs.

Application 15–1

Tell whether an informative or interpretive report should be written to develop each of the following. Review the kinds of informative and interpretive reports in this and the preceding chapter.

1. A report forecasting the weather.
2. A report that concludes that caffeine contributes to the growth of cancer.
3. A final stock market report for a certain day.
4. A profit and loss report for a small restaurant for the past six months.
5. A letter to a superintendent of schools explaining the amount of work accomplished on laying the foundation for a new high school in your city since the last report.
6. A letter report explaining to what degree an air pollution device installed in a steel mill for which you work is effective.
7. A report identifying and describing the pick-up trucks in use at the present time for a bakery in your city.
8. An insurance claim report that explains what the writer deduced as the causes for the accident.
9. A report of your employer's losses from equipment failure during a certain period of time.
10. A report explaining what you consider to be the causes for the increase in the cost of a certain product that you recently purchased.

Application 15–2

Using the list below as a guide, try to classify each of the report descriptions. It may be helpful to review the reports studied in the preceding chapters.

a. Status Reports
b. Progress Reports: Preliminary, Interim, Final
c. Narrative—Historical
d. Narrative—Process or Procedural
e. Interpretive—Evaluation
f. Interpretive—Problem: causal, proposal, or feasibility
g. None of the above

1. A report that resulted from a close observation and examination of the operation of a short-order restaurant to determine why it was operating at a loss or profit.
2. A report distributed by a manufacturer to a wholesaler, urging him to purchase a product in carload lots to enable him to reduce the cost of the merchandise by 10 percent.
3. A report summarizing the kinds and numbers of various crimes committed in a certain city within a specific period.
4. A report that stipulates the regulations to which an employee must adhere to avoid being discharged.
5. A report giving the events occurring in a person's life in the order in which they occurred so that the appropriate action may be taken in handling his case.
6. A report in the form of a letter written to the editor of a newspaper, containing the causes and reasons in support of the residency requirements for voting in a certain city.
7. A complaint report filed with the local police, urging that an arrest warrant be issued for a certain person.
8. A report that describes the condition of a patient on the date of admission, the changes that have occurred since admission, the problems or symptoms that may have occurred during the treatment, and the foreseen changes that should take place in the future if the prescribed treatment is followed.
9. A report tracing the sequence of happenings that took place during a traffic accident or some other mishap.
10. A short instruction manual written to assist practical nurses in learning the procedure for changing a bed while the patient remains in it.
11. A report written by a waitress as a part of a claim file, explaining in time sequence what happened when she spilled a tray of food on a customer.
12. A description of the main dishes that are usually served from the inflight kitchen during a certain flight.
13. A manual to be used in training air stewardesses how to prepare a tray of food, how to carry it to avoid spillage, and how to serve it.
14. A pilot's report telling where his flight originated, the time and locations of his inflight landings, the location at which he is writing this entry, the remaining stopovers, and the time he expects to reach his destination.
15. A description of a construction job before the work has started, to aid your employer to determine the number of workers to assign to the project.

16. A special report to show your employer that it is possible ten years from a certain day to consolidate the offices in a certain state by using an answering service.
17. A report judging various kinds of bond paper and concluding that the paper from a specific paper supply distributor is the most suitable for your company stationary.
18. A report resulting from the interview with an employee injured and the inspection of the tools and apparatus involved in a mishap occurring on an assembly line to make recommendations as to how the same accident can be avoided in the future.
19. A report tracing the origin and development of a certain component of a computer to enable the computer to be effectively used in a specific application.
20. A report describing the breakdown of a computer and the indications that lead to the conclusion that a specific component was inadequate for the amount of use of the computer during busy hours.
21. A report from the head sales clerk of a certain department-store counter giving information about the inventory of certain merchandise, intended to enable the receiver to forecast sales so that an order can be placed with a supplier.
22. A report describing the condition of a machine in a department so that someone can judge when replacement will be needed.
23. A report from the accounting department indicating whether or not the company will find it economically profitable to move a plant to a new location.
24. A report resulting from inspection of a transistor from a television set and giving some causes that may have contributed to its breakdown.
25. A report explaining why solar rays have a certain effect upon human skin.

Application 15–3

On a sheet of paper answer the questions listed below. Pretend you must write a report to a group of people whom you want to persuade to invest in a hardware or some other kind of store you plan to open. First tell whether you would write an informative or an interpretive report. Second, tell which kind of either type you would write. Finally, write the report in full, and in the margins identify each main section, as done for the illustrations in this and the preceding chapter.

1. Are there other hardware businesses, and if so, how many, between the prospective location and the most highly populated area? _____Stores.
2. Is this spot the most convenient hardware-store location in the area? _____Yes. _____No.
3. How many other hardware stores are in this trading area? _____Stores.
4. How many of them will compete with you for hardware customers? _____Stores.
5. Do they have better parking facilities? _____Yes. _____No.
6. Do they offer the same type of merchandise? _____Yes. _____No.
7. Do you consider them more aggressive or less aggressive than your own operation will be? _____More. _____Less.

Feasibility reports

8. What other competing stores, such as variety stores that carry hardware and household items, are planned in the near future? __________.
9. Are other potential sites that are closer to the majority of customers likely to be developed in the near future? _____Yes. _____No.
10. Are your major competitors well-known, well-advertised stores? _____Yes. _____No.
11. Is there actually a need for another hardware store in the area? _____Yes. _____No.
12. How well are the hardware demands being met? _____Good. _____Fair.
13. If there are empty stores or vacant lots near the location, what is planned for them? A hardware store? _____Yes. _____No. A store that will handle hardware and household items? _____Yes. _____No.

Application 15–4

Read the paragraph below, study the drawing, and write a complete problem report. Be sure to tell specifically which kind of report you intended to write. Identify the main divisions of the report in the right margin opposite each. Use the report illustrations in this chapter as a guide. Write an abstract for your report.

Distillation is one of the oldest ways known of separating fresh water from a salt-water solution. When salt water is boiled, the dissolved salt remains behind as the fresh-water vapor is boiled away. In a distillation process, water is first boiled and then the steam, or water vapor, is cooled. This cooling condenses the steam into water again. See Figure 15.3. Thus, distillation involves adding heat energy to salt water in order to vaporize the water and then removing the heat energy from the steam to condense it into fresh water.[3]

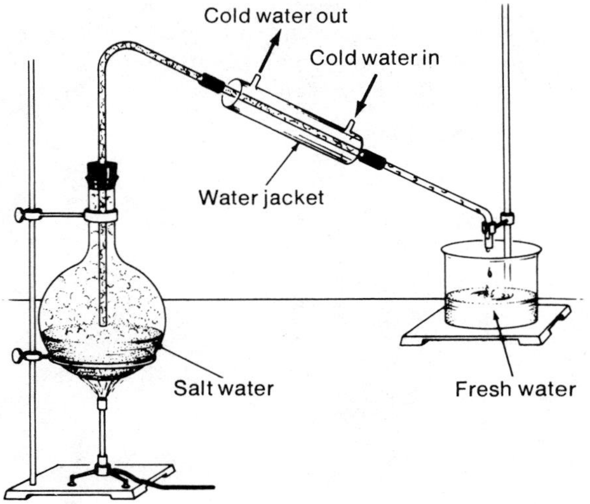

Figure 15.3
Simplified distillation setup.

[3] U.S. Department of the Interior, Office of Saline Water, "The A-B-Seas of Desalting" (Washington, D.C.: U.S. Government Printing Office, 1966), p. 3.

Application 15–5 Select one of the following and write an evaluation report of the person specified. In the right margin, identify the main parts. Be sure to write this report in paragraphs, not as a form to be completed for a job application.

1. Imagine that you are the person who does the hiring of waitresses for a restaurant chain. Describe an applicant for a waitress job. Be sure to give the physical characteristics, e.g., height, weight, etc. Give her background information, including her formal education, occupational training, amount of restaurant experience, past employment, and references. Write a brief summary of the specific job she will be expected to do. Also, indicate the person under whom she will work.
2. Imagine that you are a department head in a large general office that employees 30 people, including secretaries, general typists, billing clerks, and filing clerks. Describe the status of an employee who has applied for the position of secretary-receptionist. Give her physical qualifications, age, occupational and education background. Describe her personality and disposition since they are important qualities of a receptionist.

Application 15–6 Write a causal report analyzing a problem with which you are very familiar, perhaps a school registration problem or one related to the parking facilities. Follow the format for causal reports discussed in this chapter. Be sure to write an abstract for it. In the right margins identify each part of the report.

Application 15–7 Review the report you wrote for the preceding application and write a proposal report recommending a solution for the same problem. Also, identify the parts of this report in the margins.

Application 15–8 Write a feasibility report for a problem that may or may not become a reality in the future but for which you think contingency plans should be made. The problem you select should be within your qualifications, perhaps one related to those you wrote about in the preceding applications. Again, be sure to identify the parts of the feasibility report in the right margins. Also, write an abstract.

16

Grammar review for technical communication

Like any other craftspeople, technical writers must learn what their tools are and the job each performs. This chapter provides a review of grammar and usage to help you in technical communication. Basically we will examine the various parts of speech, and see how they function in a sentence.

PARTS OF SPEECH

The part of speech a word is cannot really be determined until we see what it does in a sentence. Although the man or woman standing next to you in a public place may look like a doctor, you can't be sure until you see him or her treating patients. Can you tell what the word "tear" means when not in a sentence? Is it a noun or verb? It may be either. For example:

> *Tear* a piece of the canvas. (a verb)
> A *tear* is secreted by the lachrymal gland. (a noun)

Learning about the functions words perform in sentences will enable you to tell what parts of speech they are. In this section we review the following eight parts of speech: verbs, nouns, pronouns, adjectives, adverbs, prepositions, conjunctions, and interjections. A brief definition is given for each, and examples illustrate what they do in a sentence.

Verbs

Verbs express action, condition, or equivalence about their subjects. Every verb must have a stated or implied subject, and every sentence must contain a verb.

> The carpenter *measured* and *sawed.* (action)
>
> She *cut* the leather and *fitted* it. (action)
>
> The weather *was* right, the fields *were* ready. (condition) (These verbs connect a condition to their subjects.)
>
> The plumber *was* the foreman. (equivalence) (The verb connects words meaning the same person, "plumber" and "foreman.")

The verbs in the English language have three principal parts: the present tense, the past tense, and the past participle (used with *have* or *has*). Most verbs are regular and should give you no trouble; their past tense is formed just by adding *ed* to the present tense:

Present Tense	*Past Tense*	*Past Participle*
talk	talked	talked
weld	welded	welded
hammer	hammered	hammered

The verbs that give trouble are the *irregular* verbs. Their correct forms must be learned, because the spelling changes are tricky and there are no rules to go by. Some verbs have an alternate spelling. Study the following list, and refer to it whenever you are in doubt.

Principal Parts of Irregular Verbs

Present	*Past*	*Past Participle*
am (be)	was	been
arise	arose	arisen
awake	awoke	awaked
bear (= carry)	bore	borne
beat	beat	beaten
become	became	become
begin	began	begun
bend	bent	bent
bid	bid	bid
bind	bound	bound
bite	bit	bitten
bleed	bled	bled
blow	blew	blown

Present	*Past*	*Past Participle*
break	broke	broken
breed	bred	bred
bring	brought	brought
build	built	built
burst	burst	burst
buy	bought	bought
catch	caught	caught
choose	chose	chosen
cling	clung	clung
come	came	come
cost	cost	cost
creep	crept	crept
cut	cut	cut
deal	dealt	dealt
dig	dug	dug
dive	dived, dove	dived
do	did	done
draw	drew	drawn
dream	dreamed, dreamt	dreamed, dreamt
drink	drank	drunk
drive	drove	driven
eat	ate	eaten
fall	fell	fallen
feed	fed	fed
fight	fought	fought
find	found	found
flee	fled	fled
fling	flung	flung
fly	flew	flown
forbid	forbade	forbidden
forget	forgot	forgotten
freeze	froze	frozen
get	got	got, gotten
give	gave	given
go	went	gone
grind	ground	ground
grow	grew	grown
hang (= suspend)	hung	hung
have	had	had
hear	heard	heard
hide	hid	hidden
hold	held	held
keep	kept	kept
know	knew	known
lay (= put, place)	laid	laid
lead (pronounced *leed*)	led	led
leave	left	left
lend	lent	lent
let	let	let
lie (= recline)	lay	lain
light	lighted, lit	lighted, lit
lose	lost	lost
make	made	made
mean	meant	meant

Present	*Past*	*Past Participle*
meet	met	met
mow	mowed	mowed, mown
pay	paid	paid
prove	proved	proved, proven
read (pronounced *reed*)	read (pronounced *red*)	read (pronounced *red*)
ride	rode	ridden
ring	rang	rung
rise	rose	risen
run	ran	run
say	said	said
see	saw	seen
seek	sought	sought
sell	sold	sold
send	sent	sent
set (= place)	set	set
sew	sewed	sewed, sewn
shake	shook	shaken
shave	shaved	shaved, shaven
shine	shone, shined	shone, shined
shoot	shot	shot
show	showed	shown, showed
shrink	shrank	shrunk
sing	sang	sung
sink	sank	sunk
sit	sat	sat
slide	slid	slid
sling	slung	slung
sow	sowed	sown, sowed
speak	spoke	spoken
speed	sped	sped
spin	spun	spun
spring	sprang	sprung
stand	stood	stood
stick	stuck	stuck
sting	stung	stung
strike	struck	struck, stricken
string	strung	strung
strive	strove	striven
swear	swore	sworn
sweep	swept	swept
swim	swam	swum
swing	swung	swung
take	took	taken
teach	taught	taught
tear	tore	torn
tell	told	told
think	thought	thought
throw	threw	thrown
wear	wore	worn
wind	wound	wound
wring	wrung	wrung
write	wrote	written

Verbals It is important that you not confuse verbs with verbals. Verbals look like verbs but perform mainly as nouns, adjectives, or adverbs. There are only three kinds of verbals: infinitives, gerunds, and participles.

1. The infinitive usually is preceded by "to" and does the job of a noun, adjective, or adverb.

   To learn diamond cutting was her dream.
   (noun, subject)
   She tried *to learn* diamond cutting.
   (noun, object)
   She worked *to learn* diamond cutting.
   (adverb, modifies verb worked)
   She had a real urge *to learn* diamond cutting.
   (adjective, modifies noun "urge")

2. The gerund ends in *ing* and acts as a noun.
   The *grinding* in the engine was loud.
   (subject of "was")
   He heard the *grinding*.
   (object of the verb "heard")
   The mechanic stopped the engine from *grinding*.
   (object of the preposition "from")

3. The participles have two forms, past or present. They act as adjectives.
   Present participles look like present-tense verbs + *ing*.

   a *working* man ("working" modifies the noun "man")

   Past participles, the third principal part of verbs, are used as verbs only when they have helping verbs; otherwise, they act as adjectives.

   a *burned* (or burnt) building
   a *dyed* piece of cloth
   the money *spent*

The following are sentence fragments, not complete sentences, even though the first word is capitalized and a word looking like a verb is in each:

Working hard and getting things done.

Earning a living.

To invest in a business when the market is right.

Here are the same verbal phrases changed into complete sentences:

Working hard and getting things done, he felt better.

Earning a living isn't always easy.

To invest in a business when the market is right may be profitable.

Noun A noun names what the verb is saying something about. The noun serving as the subject of the verb may be the name of anything, a person, place, idea, or object. The subject noun (or its substitute) and the verb form the core of a sentence. Without both a subject and a verb (stated or implied), you can't have a complete sentence.

> *Solomon* drove the *truck* into the *garage*.
>
> Good *salesmanship* increases *profit*.
>
> The *clerks* in the *store* sold the *dresses*.
>
> Paint the *house*. (The subject "you" is implied.)

Pronoun A pronoun stands for or takes the place of a previously mentioned noun. In the following sentences the pronouns stand for their respective nouns in the preceding examples.

> *He* drove *it* into the garage.
>
> *It* increases profit.
>
> *They* sold *them*.
>
> Paint *it*.

Adjective An adjective limits or describes (modifies) a noun or pronoun.

> limiters: *every* employee, *that* hammer, *most* machines
>
> describers: *hard* metal, *rough* wood, *good* worker

Adverb An adverb limits or describes (modifies) a verb, an adjective, or another adverb.

> verb modifier: worked *hard*, pounded *rapidly*, paid *promptly*
>
> adverb modifier: drilled *too* long, pumped *very* softly
>
> adjective modifier: *excessively* high temperature, *very* strong steel

Preposition A preposition establishes a relationship between a noun (or a following pronoun) and another word in the sentence.

> pay *with* cash
> (*with* relates *cash* to *pay*)
>
> went *into* a business *with* a future
> (*into* relates *business* to *went*, and
> *with* relates *future* to *business*)

Coordinate Conjunction

1. Connects equal words:
   sand *and* gravel
   cash *or* other assets
   not mortar *but* cement
2. Connects equal phrases:
   They went into business *and* into debt.
   She worked not in the office *but* in the garage.
3. Connects equal clauses:
   Either you buy now, *or* you pay a higher price.
   I invested, *for* I saw the bull market coming.
   I was a mechanic, *and* I knew the car was defective.

Subordinating Conjunction

Subordinating conjunctions join dependent clauses—acting as nouns, adjectives, or adverbs—with independent clauses.

1. As a noun (*whether, what, that*)
   I'll decide *whether he can do the job.*
   (object of verb "decide")
   What the machine can produce is known.
   (subject of verb "is")
   The agent knew *that the train would arrive.*
   (object of verb "knew")
2. As adjective (most dependent clauses are introduced by relative pronouns, see p. 287)
   He did not know the hour *when we arrived.*
   That is the house *where Henry Ford was born.*
   The time *when I saw him* is forgotten.
3. As adverb (*while, since, because, although, until, when, then, before*)
   Cost of goods increases *while inflation lasts.*
   Prices were lowered *because they were not competitive.*
   The company has been closed *since the strike began.*

Interjection

Interjections express emotions or strong feelings.

My, they had a good business.

Gosh! I forgot to turn the power off.

Oh, if we only had more cash.

Help!

Yes (or No), I'll repair your roof.

PARTS OF A SENTENCE

Once again, it is important to bear in mind that you can't tell what a word is from its appearance, but only by the way it actually functions in a sentence, that is, whether it serves as subject, verb, object, and so forth. In the following examples notice how the word "cost" and its derivatives perform different roles in each sentence.

> The *cost* of the car is high.
> (noun because it is the subject of the verb "is")
>
> The car *cost* too much.
> (verb with its subject "car")
>
> It is the *cost* I'm worried about.
> (noun because it is an equivalent word meaning the same thing as "it")
>
> I gave the *cost* to them.
> (noun because it is the object of the verb "gave")
>
> *Costing* too much, the car was not sold.
> (participle modifying the noun "car")
>
> I couldn't believe the car's *costing* so much.
> (gerund acting as the object of the verb "believe")
>
> I didn't want the car to *cost* very much.
> (infinitive acting as an adjective modifying the noun "car")

Every sentence has two main parts, a *subject* and a *predicate verb*. Most sentences have one more part, a completer called a *complement*. The predicate verb says the main thing about its subject. When a complement is needed, it completes the meaning of the predicate verb.

Predicate Verb A verb becomes a predicate verb when it is joined to a subject in a sentence, and together they make the spinal column of the sentence.

> Plastic *melts.*
>
> Fiber glass *insulates.*

Subject Except in questions, the subject of a predicate verb usually, not always, comes before it. When the subject does the action, the verb is said to be active voiced. But when the subject receives the action, the verb is called passive voiced. The subject most often is a noun or pronoun.

> He melted the plastic and poured. (active voiced)
> The plastic was melted and poured by him. (passive voiced)

Complement A complement is a word, phrase, or clause needed to complete the meaning intended by the subject and predicate verb. Not all subjects and verbs need completers, but most do. Sometimes a predicate adverb is used.

John worked.
(This doesn't need any more, if that's all the writer intended to say. But what if he or she wanted to say more—for example, John worked a machine, a crew of men, a pump handle . . .?)

John worked hard.
("Hard" is not a completer because it just modifies the word "worked." It is a predicate adverb. It doesn't complete the meaning of the subject and verb.)

There are several kinds of completers: direct object, indirect object, and predicate noun or adjective.

"Arnold sent . . ." (action not complete)

1. Direct object:
   Arnold sent the *order*.
   Arnold sent the *invoice*.
2. Combination of direct and indirect object. (An indirect object names the person to or for whom an action is taken.)
   The cashier sent *him* the bill.
   The florist sent *Mary* an application.
3. Predicate noun or adjective: Expresses condition or equivalence to complete the meaning of the subject and verb.
   The plumber was *skillful*.
   (predicate adjective)
   The supervisor was *Harriet*.
   (predicate noun expressing equivalence, the same thing; "supervisor" = "Harriet")
   They were the *workmen*.
   (predicate noun "they" = "workmen")

It is not correct to use an adverb in the place of a predicate adjective. When you can substitute a form of the verb "be" (is, am, are, was, were) for a "sense" verb such as feel, look, smell, taste, the verb before it is a linking verb and needs a predicate adjective, not an adverb, to complete its meaning.

Not correct:	The secretary felt miserably about the error. ("Was" can be inserted for "felt.")
Correct:	The secretary felt miserable about the error.
Not correct:	The engine sounds smoothly. ("Is" can be substituted for "sounds.")
Correct:	The engine sounds smooth.

Modifier Modifiers are words or word groups that describe, point out, or limit another part of a sentence. Adjectival and adverbial elements are the two main kinds of modifiers. The adjectival modifiers are those that limit nouns or pronouns. Adverbial elements modify verbs most of the time, but often modify adjectives and other adverbs.

The durable steel was tested extensively.
("Durable" is an adjective modifying the noun "steel."
"Extensively is an adverb modifying the verb "was tested.")

Some of the following phrases and clauses will illustrate groups of words used as adjectives or adverbs.

Phrase A phrase is a group of words without a subject and verb and acting as a part of a sentence.

VERB PHRASE: A verb phrase is a predicate verb consisting of more than one word.

The house *was built* by that contractor.
That item *has been sold.*
The problem *is being investigated* for the owner.

VERBAL PHRASE: A verbal phrase is one beginning with a participle, gerund, or infinitive.

Needing the raw material, we signed the order.
(Participle "needing" modifies pronoun "we"—adjective.)

Elmer Hahn ordered the machine, *knowing its capacity.*
(Participle "knowing" modifies noun "Elmer Hahn"—adjective.)

Signing the contract is scheduled for today.
(Gerund "signing" is subject of "is scheduled"—noun.)

He opposed *selling the business.*
(Gerund "selling" is object of verb "opposed"—noun.)

The manager wrote *to inform them.*
(Infinitive "to inform" modifies verb "wrote"—adverb.)

She hurried *to attend the conference.*
(Infinitive "to attend" modifies verb "hurried"—adverb.)

PREPOSITIONAL PHRASE: Prepositional phrases express relationships between nouns or pronouns after them (objects) and other words. (Refer to the explanation of preposition as a part of speech.)

The electrician went *into business for himself.*
(Preposition "into" links "business" to "went"—adverb.
Preposition "for" links "himself" to "business"—adjective.)

The store *with the merchandise* had a sale.
(Preposition "with" links "merchandise" to "store"—adjective.)

Clause Clauses are groups of words that do have a subject and predicate. There are two kinds of clauses: dependent and independent.

INDEPENDENT CLAUSE: An independent clause is one that by itself can make a complete sentence. Two or more are often joined by coordinating conjunctions to make compound sentences.

She watched the customer carefully. She knew when to try to sell him the car.
(Two simple sentences, each with one independent clause.)

She watched the customer carefully; she knew when to try to sell him the car.
(One compound sentence with two independent clauses joined by a semicolon.)

She watched the customer carefully, and she knew when to try to sell him the car.
(One compound sentence with two independent clauses joined by a comma and a coordinate conjunction, "and.")

DEPENDENT CLAUSE: Dependent (or subordinate) clauses act as nouns, adjectives, adverbs. They are introduced by a word that makes them depend on and serve a part of another clause. Unlike the independent clauses, they can't stand alone to make a complete sentence.

1. Adjective clause: modifies a noun or pronoun in another clause and usually begins with a relative pronoun, (who, that, which, what, etc.).

   He didn't see the driver *who drove the truck.*
   (Relative pronoun "who" introduces adjective clause modifying and dependent on "driver" in the other clause.)

   The company *that built the tool* guarantees it.
   (Relative pronoun "that" introduces adjective clause modifying "company.")

2. Adverbial clause: acts as an adverb, modifying a verb, adjective, or an adverb in another clause.

   The officer shouted *when he saw the crash.*
   (Introduced by adverbial conjunction "when" and modifies "shouted" in the other clause.)

3. Noun clause: acts as the subject or object of a verb or preposition in another clause.

   I didn't know *who would be promoted.*
   (Introduced by relative pronoun "who" and acts as the object of the verb "know.")

   The officer spoke to *whoever was there.*
   (Object of the proposition "to.")

COMMON WRITING PROBLEMS

The main purpose of this chapter is to help you avoid the common types of sentence writing errors. Sentences have to be written carefully. Writing fragments of sentences as if they were complete sentences is an error that often occurs in student writing. Other common errors are writing two independent clauses together with nothing connecting them, and joining independent clauses with only a comma.

Learning to use some simple punctuation marks will help you avoid these common types of sentence errors. You should be able to place the right punctuation at the end of your sentences and within them to help you express exactly what you mean.

Writing Whole Sentences Correctly

Sentences start with a capital letter and end with a period, question mark, or exclamation mark.

> Mechanics must be licensed in some states.
> Should mechanics be licensed?
> What next—a bill to license mechanics!

FRAGMENT: A fragment is a piece of a sentence punctuated as a whole one.

> The carpenter worked long hours all summer. *Trying to save money for the winter.*
> (a piece of a sentence consisting of a verbal phrase)
>
> *With many store clerks quitting cigarette smoking.* I think smoking in the store should be prohibited.
> (a fragment caused by separating a prepositional phrase from the clause to which it belongs)

The two preceding phrases belong with their clauses. They should have been separated only with a comma, not a period:

> The carpenter worked long hours all summer, trying to save money for the winter.
>
> With many store clerks quitting cigarette smoking, I think smoking in the store should be prohibited.

The following fragments are dependent clauses punctuated as if they were complete sentences:

> *The mason who did the brickwork.* He charged too much. That is the merchant. *Who bought the ad.*

These should have been written as follows:

> The mason who did the brickwork charged too much.
> That is the merchant who bought the ad.

Following are fragments that result from having important parts missing:

> Fragment: Two crates, five paper cartons, and six envelopes.
> Corrected: I delivered two crates, five paper cartons, and six envelopes. (Missing subject and verb inserted.)
>
> Fragment: Tore the envelope, opened the note, found nothing.
> Corrected: The cashier tore the envelope, opened the note, and found nothing.
> (Missing subject added.)

COMMA SPLICE: Unlike fragments, comma splices consist of two independent clauses written with only a comma between them.

> It was time to send out invoices, I got them ready.
>
> The electrician carefully wrapped the tape around the wire, with a quick jerk, he tore it.

FUSED SENTENCES: Unlike comma splices, these are independent clauses without even a comma separating them.

> It was time to send out invoices I got them ready.
>
> The electrician carefully wrapped the tape around the wire with a quick jerk he tore it.

Following are the corrections for both the preceding comma splices and fused sentences. The best way to correct them depends upon what you are trying to say, and the way you want to say it.

> It was time to send out invoices; I got them ready.
>
> It was time to send out invoices, and I got them ready.
>
> It was time to send out invoices. I got them ready.
>
> The electrician carefully wrapped the tape around the wire; with a quick jerk, he tore it.
>
> The electrician carefully wrapped the tape around the wire, and with a quick jerk, he tore it.
>
> The electrician carefully wrapped the tape around the wire. With a quick jerk, he tore it.

Avoiding Errors Within Sentences

Most of the principles related to grammar and clear constructions inside a sentence give you little trouble, but a few present difficulties. Your main problems are: (1) agreement, (2) appropriate changes in word forms, (3) modifier problems, (4) internal punctuation, and (5) capitalization.

AGREEMENT

1. Agreement of subject and verb: A verb should agree with its subject in person and number.

   I am (first person singular)
   We are (first person plural)
   You are (second person singular and plural)
   He, she, it, is (third person singular)
   They are (third person plural)

2. Subject nouns and pronouns joined by "and" take plural verbs.

   The jeweler and the doctor are friends.
   The automobile frame and engine are loaded.

3. Subject nouns and pronouns joined by or, nor, either—or, or neither—nor take verbs that agree with the nearer one.

   The pilot or the service people were angry.

   The waiters or the cook was confused.

   Neither the service people nor the pilot was angry.

   Neither the cook nor the waiters were confused.

4. Collective nouns (jury, team, herd) need singular verbs when considered as a unit, but they need plural verbs when they are used to mean more than one separate units.

   The jury was selected.

   The jury are casting their votes.
   (To avoid this awkward construction with collective nouns, you should write "the *members* of the jury . . .")

   The sales team is on the job.

   The members of the sales team were eating lunch.

5. Compound nouns meaning a single unit take singular verbs.

   Corned beef and cabbage is a tasty restaurant meal.

   Ham and eggs was the lunch special.

6. Pronouns should agree with their antecedents in number.

   Every salesperson must bring her (or his, not their) books.

   Neither of the registers had its (not their) tapes read.

7. Verbs with relative pronouns (who, whom, that, which, etc.) as subjects agree in number with the word for which the pronoun stands, its antecedent.

   The employees who are responsible will be fined.

   The employee who is responsible will be fined.

8. Avoid confusing reference of pronouns to antecedents.

Not clear:	The owner asked the cashier to count his money. (Whose money?)
Better:	The owner asked the cashier to count the cashier's money.
Not clear:	The engine made such a noise it embarrassed me. ("It" seems to refer to "noise" but is intended to refer to the whole idea of the noise made by the engine.)
Better:	The engine made such a noise I was embarrassed.
Not clear:	She hated working the press, that made the job hard. ("That" seems to refer to "working", but it is meant to refer to the whole idea of hate for working the press.)
Better:	Because she hated working the press, the job was hard.

APPROPRIATE CHANGES IN WORD FORMS

To show how they are related to other parts of a sentence and to indicate their functions, some words change in form, in their spelling. You have seen already how nouns and pronouns change to indicate whether they are singular or plural and whether each is first, second, or third person. To indicate what they do in a sentence, pronouns change their form. Verbs change theirs, depending upon which tense, person, number, and voice they are intended to express. Even adjectives and adverbs change in form to indicate the degree of comparison they make in a sentence. To write clearly and accurately, you have to learn about these changes in the case forms of pronouns.

Case Forms Some pronouns change in spelling, depending upon whether they serve as subjects or objects. These different forms, ways of spelling, are cases. Pronouns may change in form also to show whether they are singular or plural and to show ownership. Nouns used to do the same as pronouns, but now they just change to show plurals and possessive form.

Nominative (Subjective)	I	you	he, she	we, they
Possessive	my, mine	your, yours	his, her, hers	our, ours
Objective	me	you	him, her	us, them

Pronouns in the nominative (or subjective) case are used mainly as:

1. Subject.
2. Predicate nominative (an equivalent, just another word for the subject).

She threaded the lead pipe.
(Subject of verb.)

The plumber is she with the pipe.
(Predicate nominative.)

The objective case is mainly used when the pronoun is:

1. Direct or indirect object of a verb.
2. Object of a preposition.

The manufacturer's representative sold the machine to *them.*
("Them" is the object of the preposition "to.")

The representative convinced *them.*
(Direct object.)

The representative sold *them* the machine.
("Machine" is the direct object and "them" is now the indirect object.)

The possessive form of pronouns is used to show ownership. When the possessive pronoun modifies a noun, it is really a pronominal adjective.

Pronoun:	Hers is repaired. Ours is filled.
Pronominal adjective:	Her wrench is repaired. Our order is filled.

Plural Noun Forms

The plurals of most nouns are formed by adding *s.* When the noun ends in *s, sh, ch, x* or *z,* the plural is formed by adding *es* to facilitate pronunciation. Some nouns don't change in form whether they are singular or plural.

nail – nails
casting – castings
trap – traps
deer – deer
sheep – sheep

truss – trusses
crutch – crutches
swish – swishes
fuzz – fuzzes

Some plurals are formed in an unusual way:

child – children
ox – oxen
half – halves
lady – ladies
man – men

Verb Forms

Verbs along with the other important things they do, are timing devices within sentences. To tell when something exists or happens, the verb or

Appropriate changes in word forms

one of its helpers has a certain form. Using the wrong form gives the reader the wrong time, a common writing error.

Our verbs have six main tenses—three simple tenses and three perfect tenses. Simple tenses express more general present, past, and future time. Perfect tenses state or imply a time before which something exists or happens.

SIMPLE TENSES

Present: I bathe the patient.
Past: I bathed the patient.
Future: I will (or shall) bathe the patient.

PERFECT TENSES

Present perfect: I have (he, she has) bathed the patient. (in the near past)
Past perfect: I had bathed the patient before the doctor's arrival. (before something happened)
Future perfect: I will (or shall) have bathed the patient by visiting hours. (before some future event)

Progressive Verb Forms

Progressive verb forms always have helping verbs to tell when the event is in progress, and the helper indicates the tense by changes in its form.

SIMPLE PROGRESSIVE

Present: I am welding.
Past: I was welding.
Future: I will be welding.

PERFECT PROGRESSIVE

Present perfect: I have (or he or she has) been welding until now.
Past perfect: You had been welding until my arrival.
Future perfect: I will (or shall) have been welding four hours when you arrive.

Emphatic Verb Form

The emphatic form of verbs is used to emphasize that something is or was done. The helpers, *do* or *did*, indicate the tense. As in the progressive form, the main verb remains the same.

Present: I do work.
Past: I did work.

Some Common Verb and Verbal Errors

Verbs in the same sentence or the same sequence of sentences should have logical time relationships.

Incorrect:	They *melted* the metal and *pour* it into molds.
Correct:	They *melted* the metal and *poured* it into molds.
Or:	They *melt* metal and *pour* it into molds.

Incorrect:	When the officer *stopped* me, he *gives* me a ticket.
Correct:	When the officer *stopped* me, he *gave* me a ticket.
Or:	When the officer *stops* me, he *gives* me a ticket.

When you can, use the active voice of an action verb, not the weak passive voice. As stated before, the active voice is when the subject of the verb performs the action, and the passive voice is used to indicate the subject is acted upon. The tense of the passive verb is told by a weak helper; therefore, the whole verb isn't as forceful as one in the active voice. Also, notice the weak preposition added to make a sentence in the passive voice.

Active voice:	I sold the house.
Passive voice:	The house *was* sold *by* me.
Active voice:	John will prepare the menu.
Passive voice:	The menu will *be* prepared *by* John.

MODIFIER PROBLEMS

Dangling Modifiers

Danglers most often result when words that look like verbs, but act as adjectives (participles), do not clearly refer to the words they modify. This causes them to dangle. Refer to the discussion on pages 281 and 286 before looking at the following examples.

Dangles:	Grinding and squeaking, I heard the pump start. (The participles, "Grinding" and "squeaking," appear to be describing "I". They are dangling because their connection to "pump" is not clear.)
Better:	Grinding and squeaking, the pump started.
Dangles:	Setting off the dynamite, a hole was blasted. (Did the hole set off the dynamite?)
Better:	Setting off the dynamite, the engineer blasted a hole.

Comparative Forms of Adjectives and Adverbs

Adjectives and adverbs are often used to express any of three degrees of comparison: the positive, the comparative, or the superlative. The positive degree simply points out a characteristic. The comparative degree is used when a comparison is made between characteristics of only two things. The superlative is used to compare more than two.

Modifier problems

The comparative degree of adjectives and adverbs is formed by adding "er" or, when "er" would make the word hard to pronounce, by using the word "more." The superlative degree is formed by adding "est" or by using "most."

Positive Degree (*no comparison*)	*Comparative Degree* (*only two*)	*Superlative Degree* (*more than two*)
strong	stronger	strongest
large	larger	largest
durable	more durable	most durable
rapidly	more rapidly	most rapidly

Insert the word "other" to make a comparison clear and accurate.

Unclear: Detroit is larger than any city in Michigan.
(Since Detroit is in Michigan and not in another state, the word "other" is needed to make a clear accurate statement.)
Better: Detroit is larger than any other city in Michigan.

Place modifiers as close as you can to the words they modify to express exactly your meaning.

Unclear: I *only* have seen the machine working a *short time.*
Better: I have seen the machine working *only a short time.*

INTERNAL PUNCTUATION

Many of the expressions we use in speaking—our voice intonations, facial expressions, and limb gestures—are conveyed in writing by punctuation marks. Your success in written technical communication will depend partly upon how well you punctuate.

Most people have little trouble with end punctuation: the period, question mark, or exclamation point. The various internal punctuation marks, however—especially the comma—require careful study. The following explanations should provide you with a basic understanding of how to use the most common types.

Apostrophe

Use the apostrophe for:

1. Possessive forms of indefinite pronouns and nouns, *but not of personal pronouns:*
   Mary's, my uncle's, someone's (but not *its, theirs, ours*)
2. Contractions or omissions:
   didn't for *did not,* won't for *will not,* we're for *we are,* '76 for 1976
3. Plurals of numbers and letters:
   Dot your *i*'s and cross your *t*'s.
   Don't use too many *you know*'s in speaking.

Comma The main purposes of commas are to separate, to set off, or to enclose.

1. To separate introductory verbal phrases (see page 286):

   To succeed in business, you should have good credit.
2. To set off introductory prepositional phrases (see page 286):

   With our inventory built up, we were ready to open.
3. To set off introductory adverbial clauses (see page 287):

   Because we expected continued inflation, we bought as much merchandise as we could afford.
4. To separate clauses joined by coordinating conjunctions (*and, but, or, for, nor*). Commas should not be used before adverbial conjunctions (*consequently, however, therefore, moreover, still, and yet.*)

   The production problem was investigated but no solutions were recommended.

   Wrong: The production problem was investigated, however, no solutions were recommended.

   Right: The production problem was investigated; however, no solutions were recommended.
5. To separate more than two items in a series, though some organizations establish a policy of omitting the comma before *and*:

   Noise in the factory was loud, discordant, and hard to endure.
6. To separate two or more adjectives that modify a noun or pronoun in somewhat the same way. To determine whether or not these adjectives are coordinate, see if the word "and" will fit smoothly between them. If the insertion of "and" will alter the meaning, don't use the comma.

   I have never met a more eager, ambitious person.

   We saw many quaint agricultural tools. (No commas)

Nonrestrictive appositives and modifiers, parenthetical expressions, mild exclamations and interjections, parts of dates and address, and persons addressed directly are set off or enclosed by commas.

1. Nonrestrictive words or groups of words merely supply unessential additional information; they do not restrict, limit, or point out exactly.

   Marshall Amorry, who waited on me, forgot to check my tires.

   (The reader can tell who did the waiting on me just by the attendant's name, without the nonrestrictive clause "who waited on me." Therefore, it should be set off by commas.)

   The gas station attendant who waited on me forgot to check my tires.

   (The reader can't tell which attendant forgot to check my tires without "who waited on me." It is needed as a restrictive clause; therefore, it should not be set off.)
2. Nonrestrictive appositives also just add additional information that does not limit or point out. It is just another noun or noun substitute that names the same thing.

Internal punctuation

John Malley, the mechanic, collects bumpers.
("John Malley" points out who without "the mechanic"; consequently, it is set off.)
The mechanic John Malley collects bumpers.
("John Malley" is needed to tell exactly who; it is restrictive and not set off.)

3. Parenthetical expressions are set off by commas because they interrupt the main line of thought. They are not an essential part of a sentence.

   The truck, according to the policeman, hit a bridge abutment.

4. Mild exclamations or interjections.

   Yes, the salary increase was refused.
   My, you are a generous employer.

5. Direct address.

   Mr. Long, have you considered my promotion?
   We ask you again, Madame President, what is your wish?

6. Words identifying the speaker in a direct quotation.

   "The glucose tolerance test," said Dr. Adams, "is not the only way to diagnose diabetes."

   "Several children accidentally die each year on playground equipment," the insurance man said.

7. Dates, places, addresses.

   On June 1, 1976, the plant located at 1637 Pine Road, Dallas, Texas, was moved to Arizona.
   (Be sure to see the commas after 1976 and after Texas.)

Semicolon

Semicolons are mainly used to separate independent clauses that are not joined by a coordinating conjunction (or, for, and, but) preceded by a comma. Also, a semicolon is used before an adverbial conjunction (consequently, therefore, however, moreover, yet, still) linking independent clauses.

The doctor called for the little patient; the boy came promptly.

She knew the problem's cause; nevertheless, she authorized the investigation.

Colon

In sentences and paragraphs, a colon is placed after an independent clause. It is also used to indicate a formal pause before a list or a quotation and to introduce an explanation.

Workers should bring the following: a tape measure, crescent wrench, hammer. and pliers.

Quotation Marks

Quotation marks are used to identify a direct quotation. Single quotation marks are used for a quotation within a quotation.

"I want you to make a thorough investigation," the inspector said.

The clerk said, "The robber shouted, 'You'll die if you move.' "

Quotation marks are also used to identify titles of short articles and reports. Titles of long articles and reports are underlined.

The short report was entitled "Hazards of Playground Equipment."

"Transportation Noise and Control" is an article discussing physical harm caused to eardrums.

Dash A dash indicates a sudden and often abrupt change in the main line of thought. Students are inclined to use dashes when they don't know what else to use.

Enough talk about bankruptcy—now let's discuss our vacations.

Parentheses Stronger than commas, parentheses are used to enclose side comments and other interrupters.

The complaining customers (and the loudest one was there) were ushered in to talk with the owner.

Underlining Underline the following: [1]

1. Titles of books, journals, magazines, newspapers, and long formal reports.
   The Wall Street Journal (newspaper)
   Rail and Water Interface at Hampton Roads (report)
   Road Construction (magazine or journal)
   Silent Spring (book)
2. Names of aircraft and ships.
   Andrea Doria, Sky Clipper
3. Foreign words.
   caveat emptor (Latin—let the buyer beware)
   modus vivendi (Latin—a way of living)
4. A reference to a specific word or letter.
   Do you spell your name with a g?
   There are too many I's in his letter.

---

[1] Note that in printed material such as books and magazine articles, italics are usually used instead of underlining.

CAPITALIZATION

Following are some principles to guide your use of capital letters. Capitalize:

1. The first word of a sentence and of a quotation that is a complete sentence.

   Here are the foreman's exact words: "If you want a job, see me tomorrow."
2. Proper nouns, those that name certain individual members of a class. Don't capitalize common nouns, those that name whole classes.

Tulsa, Oklahoma—city, state	Grace Smith—woman
Joe Louis—boxer (man)	Gratiot Road—street
August—month	Christmas—holiday

   Do not capitalize spring, summer, fall, winter when used as names of seasons.
3. Important words in titles, not less important articles, prepositions, and conjunctions unless they are the first or last words. If these less important words contain five or more letters, they may be capitalized.

   Handbook for the Home
   The Land Beyond the Mountains
4. Names of geographic areas, countries, and territories.

   Siberia
   Italy
   Upper Peninsula (Michigan)
5. Religious names and some pronouns referring to the Deity.

   Redeemer, Heavenly Spirit, God, Thine, His, Bible, Catholic, Methodist, Baptist
6. Names of family relationships when used with the person's name or when they stand for the certain person. This rule also applies to titles.

   My aunt and her son visited my Uncle Ben, but Mother didn't know Aunt Martha and her son were in town.

   The supervisor told me to report to Inspector Helms.
7. The compass points (North, South, East, West) when they refer to regions, not when they mean directions.

   Our suppliers in southern Georgia are in the South.
8. Languages and school subjects that are specifically identified by catalog number.

   I like studying business management courses, especially Business Management 102.

   Next semester, I plan to study German.

Application 16–1

How a word is used in a sentence determines which part of speech it is. Identify the part of speech each italicized word is by writing one of the following abbreviations above it.

N = noun	Adj = adjective	Conj = conjunction
Pro = pronoun	Adv = adverb	Inter = interjection
V = verb	Prep = preposition	Vbl = verbal

1. The *plumber put* some lead *on* the pipe.
2. Henry, *telephone me* your decision *immediately*.
3. Yesterday *was* the *first* day of the big *sale*.
4. Each secretary *should do* his *own* work.
5. *My*, how inflation *has increased* the cost *of* merchandise!
6. The *wrenches* were *poorly* made, *and* we couldn't use *them*.
7. The customer was *very angry because* the *needed* merchandise arrived *late*.
8. The sales manager was *happy when* he *learned of* the *new* account.
9. *Risky charge* accounts should be refused *promptly*.
10. *Going into* debt excessively is not good *for* business.
11. *To finance* a large *purchase*, you need a good *rating*.
12. *Raising* his prices, he *saved* his *faltering* business.

Application 16–2

To determine which part of a sentence a word is, you first have to discover what it does. After writing out each of the following, underline the subject with one line and the predicate verb with two lines. The subject *you* may be understood in a command or request.

1. A business should avoid excessive indebtedness.
2. A way to reduce the amount of interest is to reduce the mortgage.
3. Some advisors suggest down payments of 15 percent.
4. You should learn merchandising practices and use them wisely.
5. Good credit is essential for any business.
6. At one time, the cost for food was 5 to 10 percent below the national average in Detroit, Michigan.
7. A warranty is provided by the contractor.
8. The plans, specifications, and contract should be secured for the builder.
9. The cutter is designed to cut material such as soft wire.
10. Sidecutters are used for holding, bending, and cutting material.
11. Pocket knives are for sharpening pencils and other kinds of light cutting.
12. There are several kinds of tools used to cut, pare, and trim.
13. With less need for long term food storage, today's houses are built with recreation rooms in the basements.
14. To test a surface for flatness, carefully clean it and remove all burrs.
15. Partitions are a necessary part of a cooler.
16. Three years ago, Tiff Electronics began its search for a TV using digital design techniques.
17. Four-cycle air-cooled Stratley engines are manufactured in Bucyrus, Ohio.

18. Clutches and manual transmissions save gas.
19. The platform was made of cabinet-grade oak plywood.
20. Make a continuous spiral.
21. The legs, ends, and crossrails were oak.
22. The parts of the chandelier come in an easy to assemble kit.
23. Put the turning back and give it a thorough sanding.
24. The Moto Motor Home is built to handle crosswinds.
25. Clamp the jig to the drill-press table, bore the five holes, and insert a paper wedge to give the turning solid support.

Application 16–3

Without a completer, a sentence may not express completely what the writer intended to say. Often without one, a sentence wouldn't make sense. After writing out the sentences below, place one of the following abbreviations over the completer to indicate which kind it is. Not all of these sentences have completers. Some may have predicate adverbs to complete their meanings.

DO = direct object	P Adj = predicate adjective
IO = indirect object	P Adv = predicate adverb
PN = predicate nominative (noun or pronoun equivalent)	None = contains no completers

1. The ignition key is turning clockwise.
2. The television set's tuning knob broke.
3. Electronic flashguns may be used with color film.
4. You should sell the highest bidder the merchandise.
5. Modern copying machines have simplified the job of making duplicates.
6. The tools were packed poorly.
7. All of the ordered components arrived.
8. He painted the fender's outer surface quickly.
9. The shipper sent the customer the plastic granules.
10. The truck hit the curb and the divider.
11. The vehicle contained leather covered bucket seats.
12. Nuts and bolts attach the hinges.
13. A 3-stage transformer amplified the error pulse.
14. The wages there are better.
15. Palmer Aircraft built the plane and its equipment.
16. Run a small heat duct carefully.
17. Door handles should be treated.
18. Properly installed, a chimney may be expensive.
19. Send them the invoice.
20. Aluminum and vinyl siding have desirable characteristics.

Application 16–4

A phrase is a group of words without a verb. Beginning with a preposition or a verbal, each serves as a noun, adjective, or adverb. First tell which kind of word begins each of the following underlined phrases: preposition, infinitive, gerund, or participle. Next, tell whether each serves as a noun, adjective, or adverb. Verb phrases, a main verb with one or more helpers, are not included.

1. You should send the letter to him.
2. Ironing the shirt, the man perspired.
3. Steel is a metal requiring some scrap iron.
4. The shipment was too large for air freight.
5. There is a variety of welding devices.
6. American Aircraft built a fast plane for them.
7. Cooking short order items requires skill and patience.
8. He enjoys buying old cars and repairing them.
9. Building bridges is a demanding job.
10. The painter worked his way up the ladder.
11. The roof with the asbestos tile needed to be repaired.
12. Operating a welder requires some knowledge of metal.
13. The police officer in the car called for help.
14. The cashier went to the bank to make a deposit.
15. Working in a garage was exciting for him.

Application 16–5

A clause, unlike a phrase, is a group of words with a predicate verb. There are two kinds of clauses. The independent clause does not serve as a noun, adjective, or adverb and by capitalizing its first word and placing the proper end punctuation, it will make a complete sentence. A dependent (subordinate) clause serves as a noun, adjective, or adverb; therefore, it depends for its meaning on a word in the sentence's independent clause. It cannot stand alone and express its intended meaning by itself.

First tell whether each of the following underlined clauses is independent or dependent. If it is a dependent clause, tell whether it serves as a noun, adjective, or adverb.

1. The company which builds auto parts supplies them to auto manufacturers.
2. One company manufactures copying machines, and the other manufactures the paper for them.
3. That manufacturer pays higher wages; consequently, my friends like to work there.
4. The small factory makes kitchen cabinets when there is a large enough order for them.
5. "Industry must regulate itself," was his main argument.
6. Tell the foreman that the coal will be delivered today.
7. That the ore was radioactive was known.
8. The carpenters went home, and all the work stopped.
9. When the crew arrived, the crane was started.

10. Students who study mechanical drawing are given jobs.
11. Many colleges offer technical writing because industry demands that employees know how to communicate.
12. When business was poor, they closed the shop.

Application 16–6

Learning to use simple punctuation correctly will help you to avoid some very common sentence faults. Identify what kind of error: *fragment, comma splice,* or *fused sentences,* you find in the following, and rewrite each correctly.

1. The baker carefully carried the tray of bread reaching for the oven door, he opened it.
2. The electrician worked for three hours trying. Wanting to save the poor woman who called him to repair her expensive dishwasher.
3. The bricklayer who did the work, who was inefficient.
4. With our workers quitting. We should pay them more.
5. The bank teller was very careful with the account, she knew that her records were correct.
6. To encourage employees and that way enabling them to succeed.
7. The men did not go home, the work was not done.
8. His mother is a truck driver and his father is a waiter.
9. Harry is studying hard for the final exam, that's why he didn't come to work today.
10. You should work hard. Because it's the only way to succeed.

Application 16–7

Verbs must agree with their subjects in number. If the subject is singular, its verb has to be singular, and if the subject is plural, its verb has to be plural. Rewrite the sentences below, correcting the errors in agreement.

1. The mechanic and the doctor is a friend of the judge.
2. Neither the gas station attendant nor the restaurant owner were at that meeting.
3. Everyone must bring their salesbooks.
4. Wieners and beans are a profitable restaurant dish.
5. Either the cab driver or his fares was arrested.
6. The store clerks, not the manager, was guilty.
7. Creamed chicken on toast are good to serve for the school banquet.
8. Scissors is the only thing you need.
9. Neither the witnesses nor the victim were watching the car.
10. Either the lawyers or the judge were wrong.

Application 16–8

The case forms of pronouns depend upon whether they serve as subjects or objects of the verb or as predicate nominatives (noun or pronoun equivalent). Review the explanation, then choose the correct form.

1. Jane can work at the machine as fast as either Jim or (I, me).
2. From the evidence presented, he inferred the guilty one was (he, him).
3. The sales commission will be divided between John and (she, her).

4. The floor supervisor is (her, she) with the pencil.
5. The purchasing agent gave (they, them) the order.
6. (Him, he) over there is the boss.
7. The chairperson is (her, she) at the desk.
8. (He, him) the man unloading the truck is the owner of the store.
9. Each of the men had to show (his, theirs) card.
10. (Me, I) and my father started the business.

Application 16–9

To form the plural of most nouns add *s* or *es* if it ends in *s, x, sh,* or *ch.* To form the possessive of nouns not ending in *s* add *'s.* To those ending in *s* add only an apostrophe.

Write the correct plural and the correct singular and plural possessives for each of the following:

1. hammer
2. pliers
3. tax
4. church
5. machinist
6. vise
7. axe
8. inquiry
9. wrench
10. invoice
11. cashier
12. scissors
13. bucket
14. attorney
15. birch

Application 16–10

For each of the following, write the correct form of the predicate verb, and indicate which tense it is: present, past, future, present perfect, past perfect, or future perfect.

1. I had just ate my supper when the repair truck come.
2. During the night, several pipes had froze and bursted.
3. The tailor had shrank the dress.
4. Although I could of went to college, I'm proud of being a business man.
5. I had not spoke to the insurance man because I haven't knowed him long.
6. The nurse told the patient he must sit down because his temperature had raised again.
7. The used car dealer had gave a low price for the car.
8. He sat the adding machine on the desk.
9. The man is lying on his back under the car.
10. The plumber couldn't fix the pipe because the wrench broke.
11. Jim heard that the pipes in the house froze.
12. Raising over the tank was a cloud of black smoke.
13. The carpenter repaired the roof that blowed off the house.
14. The truck driver threw the gear into first and begun the long trip.
15. The patient drunk the medicine the nurse give him.

Application 16–11

First list the three principal parts, then write the requested form of each verb.

1. Past tense of *broke*
2. Past perfect tense of *burst*
3. Future tense of *go*

4. Future perfect tense of *do*
5. Present perfect tense of *bring*
6. Past perfect tense of *freeze*
7. Past tense of *ring*
8. Present perfect tense of *throw*
9. Past perfect tense of *tore*
10. Future perfect tense of *ride*

Application 16–12 The following sentences contain modifier errors. Correct each sentence by rewriting it. Also, tell which sentences have dangling modifiers.

1. Blowing the horn, the car ran over the hill.
2. I only thought that you, not they, were responsible for the work.
3. The steel mill is larger than any company in Indiana.
4. Of the two sales totals, his is the largest.
5. Eating my dinner, the factory whistle sounded.
6. After two years of college, my boss wouldn't give me a raise.
7. At the age of eight, my father retired from business.
8. There are two men in that business, and he is the richest.
9. Selling insurance is the most hard kind of selling.
10. Being heated to two hundred degrees, the chemist removed the metal from the furnace.

Application 16–13 Review the discussion of capitalization and punctuation in this chapter and correct each of the sentences below. Rewrite each sentence correctly and identify the kinds of errors you found by placing the appropriate abbreviations before each:

apos. = apostrophe	per. = period
cap. = capital	quot. = quotation marks
col. = colon	ques. = question mark
com. = comma	semi. = semicolon

1. You may use plywood Aluminum masonry, hardboard or vinyl for siding material.
2. When you insulate you must provide a vapor barrier.
3. No this crop is grown in the deep south.
4. He knew that father, a hard business man was a member of the republican party.
5. To go into business for himself he borrowed, later he had to sell his house to pay his debts.
6. After we had been in business for ten years we sold out.
7. They won't send the statement, said the lawyer.
8. Ms. Smith is the bookkeepers best clerk.
9. When building a new house be certain you have the right documents.
10. We have the hose in four shades tan, beige, sand, and grey.

11. We sent him Tools and their uses, a book for apprentices.
12. My lawyer told me to report to judge Allen and to pay whatever he orders.
13. The smelter was moved four blocks South but the warehouse was not moved.
14. There were three As in his initials an unusual occurrence.
15. Hands off, shouted the mechanic leaping toward the car.
16. To get the degree, Mr. Ryan completed accounting 106 and a course in Business Law, he graduated last Spring.
17. The company was started in Bucyrus, Ohio in 1917.
18. Our accountant who is a very honest man wouldn't do that.
19. These drawings contain the design of the house not the basic floor plans of the interior.
20. Leaky faucets fixtures and tanks waste gallons of water.

Appendixes

APPENDIX A

SOME OFTEN MISSPELLED WORDS

The reader's confidence in the accuracy of your writing is shaken by incorrectly spelled words. Following is a list of words commonly misspelled. Learn to spell them correctly.

accessible
accidentally
accommodate
accompanied
accomplish
accumulate
acquaintance
acquire
all right
altogether
amateur
angle
apologize
appearance
argument
asylum
athletic
beginning
believe
beneficial
benefit
benefited
benefiting
boundary
business
cabinet
certain
changeable
character
chief
coming
commit
committed
committing
comparatively
competition
compliment
conceivable
conferred
conscience
conscientious
conscious
consensus
contemptible
controlled
convenience
counterfeit
courtesy
criticism
decide
decision
definite
descend
descendant
describe
description
desirous
desperate
despicable
develop
diagonal
diaphragm
difference
different
disappear
disappoint
doesn't
during
ecstasy
elaborate
eliminate
embarrass
embarrassment
environment
equipment
equipped
escape
etiquette
exaggerate
excel
excellent
excite
excited
excitement
exciting
exist
existence
experience
fallacy
familiar
fascinate
finally
financier
fluorescent
forcible
foreign
foreigners
formulas
forty
friend
government
grammar
grievance
guarantee
hindrance
hosiery
humor
humorous
hypocrisy
image
imaginary
imagination
imagine
immediate
immediately
improvement
incandescent
indispensable
ineligible
inevitable
inflammation
innocence
inoculate
intercede
interest
irrepressible
irresistible
irridescent
its (possessive)
it's (it is)
jeopardize
judgment
knowledge
laboratory
legible
leisure
license
lieutenant
lonely
lose
losing
lovable
management
marriage
marry
mayonnaise
meant
melon
mercurochrome
mileage
mimicking
miscellaneous
mischievous
misspelled
movable
necessary
necessity
noticeable
occasion
occasionally
occur
occurred
occurrence
occurring
omitting
opinion
opportunity
panicky

Some often misspelled words

parallel
paralyze
parliament
passers-by
percolator
performance
perimeter
persistent
personnel
perspiration
persuade
picturesque
pleasant
pneumonia
possess
possessive
prejudice
principal
principle
privilege
probably
procedure
proceed
professor
proficient
propelling
prophecy (*n.*)
prophesy (*v.*)
pursue
questionnaire
quiet
receive
recommend
reconcilable
referred
resemblance
restaurant
resuscitate
rhythm
sacrilegious
sandwiches
scholastic
schedule
seize
sense
separate
separator
sergeant
sieve
silhouette
similar
sophomore
souvenir
sovereign
spaghetti
speech
stop
stopped
stopping
studied
studios
study
studying
succeed
success
successful
superintendent
supersede
surprise
symmetry
temperature
than/then
their/there

APPENDIX B

TWO-LETTER POSTAL ABBREVIATIONS

You can simplify addresses and increase the machinability of your mail by using the following abbreviations:

Alabama	AL
Alaska	AK
Arizona	AZ
Arkansas	AR
California	CA
Canal Zone	CZ
Colorado	CO
Connecticut	CT
Delaware	DE
District of Columbia	DC
Florida	FL
Georgia	GA
Guam	GU
Hawaii	HI
Idaho	ID
Illinois	IL
Indiana	IN
Iowa	IA
Kansas	KS
Kentucky	KY
Louisiana	LA
Maine	ME
Maryland	MD
Massachusetts	MA
Michigan	MI
Minnesota	MN
Mississippi	MS
Missouri	MO
Montana	MT
Nebraska	NE
Nevada	NV
New Hampshire	NH
New Jersey	NJ
New Mexico	NM
New York	NY
North Carolina	NC
North Dakota	ND
Ohio	OH
Oklahoma	OK
Oregon	OR
Pennsylvania	PA
Puerto Rico	PR
Rhode Island	RI
South Carolina	SC
South Dakota	SD
Tennessee	TN
Texas	TX

Utah	UT	Washington	WA
Vermont	VT	West Virginia	WV
Virginia	VA	Wisconsin	WI
Virgin Islands	VI	Wyoming	WY

APPENDIX C

TIPS ON THE USE OF NUMERALS

Do you know when to use numbers or when to spell them out? Readers understand numerals easier than numbers expressed by words, especially in very technical writing. In many instances, however, numbers are spelled out.

The following suggestions from the United States Government Printing Office *Style Manual* will help you choose the best form.

Numbers Expressed in Figures

A figure is used for a number of 10 or more with the exception of the first word of the sentence. Numbers under 10 are to be spelled, except for time, measurement, and money. Arabic numerals are generally preferred.

Petroleum came from 16 fields, of which eight were discovered in 1956.

That man has three suits, two pairs of shoes, and 12 pairs of socks.

Of the 13 engine producers, six were farm equipment manufacturers, six were principally engaged in the production of other types of machinery, and one was not classified in the machinery industry.

Units of measurement and time are expressed in figures.

a. Age:
   6 years old
   52 years 10 months 6 days
   a 3-year-old
b. Clock time (see also Time):
   4:30 p.m. (use thin colon)
   10 o'clock *or* 10 p.m. (*not* 10 o'clock p.m.; 2 p.m. in the afternoon; 10:00 p.m.); 12 m. (noon); 12 p.m. (midnight)
   half past 4
c. Dates:
   June 1935; June 29, 1935 (*not* June, 1935, *nor* June 29th, 1935)
   March 6 to April 15, 1935 (*not* March 6, 1935, to April 15, 1935)
d. Decimals: In text a cipher should be supplied before a decimal point if there is no unit, and ciphers should be omitted after a decimal point unless they indicate exact measurement.
   0.25 inch; 1.25 inches
   silver 0.900 fine

Tips on the use of numerals

e. Degrees, etc. (spaces omitted):
    longitude 77°04′06″ E.
    45.5° to 49.5° below zero
    an angle of 57°

f. Market quotations:
    4½-percent bonds
    Treasury bonds sell at 95

g. Mathematical expressions:
    multiplied by 3
    divided by 6

h. Measurements:
    7 meters
    about 10 yards
    8 by 12 inches
    9 bushels
    1 gallon

i. Money:
    $3.65; $0.75; 75 cents; 0.5 cent
    $3 (*not* $3.00) per 200 pounds

j. Percentage:
    12 percent; 25.5 percent; 0.5 percent (*or* one-half of 1 percent)

k. Proportion:
    1 to 4

l. Time (see also Clock time):
    6 hours 8 minutes 20 seconds

m. Unit modifiers:
    5-day week
    8-year-old wine
    8-hour day
    10-foot pole

n. Game scores:
    3 to 2 (baseball)
    7 to 6 (football), etc.

Numbers Spelled Out

Numerals are spelled out at the beginning of a sentence or head. Rephrase a sentence or head to avoid beginning with figures.

Five years ago . . . ; *not* 5 years ago . . . .

Five hundred and fifty men are employed . . . ; *not* 550 men are employed . . . .

A spelled-out number should not be repeated in figures, except in legal documents. In such instances these forms will be observed:

five (5) dollars, *not* five dollars (5)

ten dollars ($10), *not* ten ($10) dollars

Numbers mentioned in connection with serious and dignified subjects such as Executive orders, legal proclamations, and in formal writing are spelled out.

the Thirteen Original States

the Seventy-eighth Congress

Indefinite expressions are spelled out.

the seventies; the early seventies; *but* the early 1870's or 1870's
a thousand and one reasons
between two and three hundred horses; *better,* between 200 and 300 horses
midthirties
in the eighties, *not* the '80's *nor* 80's
twelvefold; fortyfold; hundredfold, twentyfold to thirtyfold
but 1 to 3 million
mid-1951
40-odd people; nine-odd people
40-plus people
100-odd people
3½-fold; 250-fold; 2.5-fold; 41-fold

For typographic appearance and easy grasp of large numbers beginning with *million,* the word *million* or *billion* is used.

The following are guides to treatment of figures as submitted in copy. If copy reads:

$12,000,000, *change to* $12 million

2,570,000,000 dollars, *change to* $2,570 million

Related numbers close together at the beginning of a sentence are treated alike.

Fifty or sixty miles away is snowclad Mount McKinley.

Round numbers are spelled out.

a hundred cows
a thousand dollars
a million and a half
two thousand million dollars
less than a million dollars

Fractions standing alone, or if followed by *of a* or *of an,* are generally spelled out.

three-fourths of an inch; *not* ¾ inch *nor* ¾ of an inch

one-half inch

one-half of a farm; *not* ½ of a farm

one-fourth inch

APPENDIX D

CLEAR, PLAIN, PERSONAL LANGUAGE

Use clear, plain, workhorse English to express ideas. Although the following suggestions are not always suitable, most of the time they will express what you really want to say best.

Instead of:	*Say:*
as from	from, since, after
as of this date	today
as per your letter	in your letter
ascertain	learn
assistance	help, aid
at all times	now
at an early date	soon
at your earliest convenience	as soon as you can
attached hereto	attached
balance of the day	remainder of the day
basis for	for
call your attention to	let you know, tell you
cognizance	awareness
commence	begin, start
commitment	promise
contribute	give
consummate	complete
data	information
demonstrates	shows
determine	decide, find out
due to the fact that	because
effectuate	bring about
employed (for used)	used
enclosed herewith, enclosed please find	enclosed is
enclosed you will find	I (we) enclose
endeavor to ascertain	try to find out
equivalent	equal
event	incident, happening
exercise care	be careful
expedite	hasten, hurry
experience has indicated that	I (we) learned
firstly	first
finalize	end, conclude, complete
for your information	to tell you
for the month of May	for May
for the reason that	since, because
forward	send
furnish	give
if doubt is entertained	if doubtful
if it is deemed satisfactory	if satisfactory

Instead of:	*Say:*
if you desire	if you wish
implement	carry out
in accordance with your wishes, in compliance with your request	as you requested
in addition to	besides
in the near future	soon
in the amount of	for
in the meantime	meanwhile
in order to	to
in regard to	about
in view of the fact that	as
in a position to	I (we) can
inadvertency	error (or mistake)
inasmuch as	since, as, because
indicate	show
initiate	begin
incapacitated	unable
in lieu of	in place of
insure	make sure
it has been noticed	I (we) have noticed
kindly advise the undersigned	please let me know
liquidate	pay off
loan (as a verb)	lend
locality	place
makes provisions for	does
meets with our approval	we approve
modification	change
it is not necessary that you	you need not
nominal	small
notwithstanding the fact that	although (or even though)
obligation	debt
optimum	best
over the signature of	signed by
an error on our part	our error
payable to your order	payable to you
pecuniary interest	financial interest
per diem	by the day
per annum	by the year
peruse	read
preclude the necessity	prevent (or shut out)
previous to (or prior to)	before
purchase	buy
pursuant to	under
quite	really, truly
rarely ever, seldom ever	rarely, seldom
remuneration	pay
render assistance	give help
respecting	about
reside	live
retain	keep
secure	get, take, obtain
seldom ever	seldom
still remains	remains

Clear, plain, personal language

Instead of:	*Say:*
subsequent to	after
sufficient	enough
terminate	end
this day in receipt of	today we received
transmit	send
under date of	on, dated
until such time as	until
utilization	use
wish to apologize	apologize

APPENDIX E

METRIC MEASUREMENTS

The United States is gradually converting to the metric system, already used in the sciences and in most countries of the world. Meter, liter, gram, and Celsius degree are the basic units of the metric system.

Meter: A little longer than a yard (about 1.1 yards).

Liter: A little larger than a quart (about 1.06 quarts).

Gram: A little more than the weight of a small wire paper clip (1/28 ounce).

Celsius degree (also called centigrade):

Celsius		*Fahrenheit*
−40°		−40°
0°	(water freezes)	32°
37°	(body temperature)	98.6°
100°	(water boils)	212°

Common prefixes

milli:	one-thousandth (0.001)	(1000 millimeters = 1 meter)
cent:	one-hundredth (0.01)	(100 centimeters = 1 meter)
kilo:	one thousand times (1000)	(1000 meters = 1 kilometer)

Other commonly used units

millimeter:	0.001	meter	= diameter of a wire paper clip
centimeter:	0.01	meter	= a little more than the width of a small wire paper clip
kilometer:	1000	meters	= about 0.6 mile (little more than half)
kilogram:	1000	grams	= about 2.2 pounds (little more than two)
milliliter:	0.001	liter	= 1/5 teaspoon

Other useful units

hectare: about 2½ acres
long-ton: 2,240 tons

COMMON MEASURES AND THEIR METRIC EQUIVALENTS

Common measure	*Equivalent*	*Common measure*	*Equivalent*
Inch	2.54 centimeters	Dry quart (U.S.)	1.101 liters
Mile	1.6093 kilometers	Quart (imperial)	1.136 liters
Foot	0.3048 meter	Gallon (U.S.)	3.785 liters
Yard	0.9144 meter	Gallon (imperial)	4.546 liters
Rod	5.029 meters	Peck (U.S.)	8.810 liters
Square inch	6.452 square centimeters	Peck (imperial)	9.092 liters
Square foot	0.0929 square meter	Bushel (U.S.)	35.24 liters
Square yard	0.836 square meter	Bushel (imperial)	36.37 liters
Square rod	25.29 square meters	Ounce, avoirdupois	28.35 grams
Acre	0.4047 hectare	Pound, avoirdupois	0.4536 kilogram
Square mile	259 hectares	Ton, long	1.0160 metric tons
Cubic inch	16.39 cubic centimeters	Ton, short	0.9072 metric ton
Cubic foot	0.0283 cubic meter	Grain	0.0648 gram
Cubic yard	0.7646 cubic meter	Ounce, troy	31.103 grams
Cord	3.625 steres	Pound, troy	0.3732 kilogram
Liquid quart (U.S.)	0.9463 liter		

Index

D

E

F

G

H

I

L

M

N

O

P

Q

R

S

T

U

V

W